NUCLEAR COMMAND AND CONTROL IN NATO

Nuclear Command and Control in NATO

Nuclear Weapons Operations and the Strategy of Flexible Response

Shaun R. Gregory
Lecturer
Department of Peace Studies
University of Bradford

First published in Great Britain 1996 by
MACMILLAN PRESS LTD
Houndmills, Basingstoke, Hampshire RG21 6XS
and London
Companies and representatives
throughout the world

A catalogue record for this book is available
from the British Library.

ISBN 0–333–64697–5

First published in the United States of America 1996 by
ST. MARTIN'S PRESS, INC.,
Scholarly and Reference Division,
175 Fifth Avenue,
New York, N.Y. 10010

ISBN 0–312–12914–9

Library of Congress Cataloging-in-Publication Data applied for

10 9 8 7 6 5 4 3 2 1
05 04 03 02 01 00 99 98 97 96

Printed and bound in Great Britain by
Antony Rowe Ltd, Chippenham, Wiltshire

For Tracey Jayne Gregory

Contents

List of Figures

Acknowledgements

I am indebted to Malcolm Dando and Paul Rogers for their support and encouragement over many years.

My thanks are also due to the following who in many diverse ways have helped me: William Arkin, Desmond Ball, Bruce Blair, Paul Bracken, William Burrows, Duncan Campbell, Malcolm Chalmers, Daniel Charles, Admiral Sir James Eberle, Lawrence Freedman, Owen Greene, William Hall, Paul Helm, Jolyon Howorth, Andrew Kelly, Stuart Kemp, Richard Ned Lebow, J. M. Legge, Milton Leitenberg, Malcolm Spaven, John Steinbruner, and Philip Weber. Responsibility for errors is, however, entirely my own.

I would also like to thank NATO and the MITRE Corporation for providing me with documentation and arranging visits, and the Center for Defense Information, Institute of Policy Studies, *New Statesman and Society* and the RAND Corporation for letting me look through relevant files.

Thanks are also due to the Charities Aid Foundation, the Rowntree Trust and the Sir Robert Menzies Fellowship for supporting parts of the research.

SHAUN R. GREGORY

List of Abbreviations

AAFCE	Allied Air Force Central Europe
ACCHAN	Allied Command Channel
ACCS	Air Command and Control System
ACE	Allied Command Europe
ACE-HIGH	ACE Troposcatter Communications Network
ACENET	ACE Communications Network
ACLANT	Allied Command Atlantic
ADGE	Air Defence Ground Environment
ADM	Atomic Demolition Munition (nuclear mine)
AEC	Atomic Energy Commission
AEGIS	Airborne Early Warning/Ground Environment Integration Segment
AEW	Airborne Early Warning
AFB	Air Force Base
AFCENT	Allied Forces Central Europe
AFNORTH	Allied Forces Northern Europe
AFSOUTH	Allied Forces Southern Europe
AMF(L)	ACE Mobile Forces (Land)
ANMCC	Alternative National Military Command Centre
AOC	Air Officer Commanding
ARP	Alert Response Plan
ASTARTE	Avion Station Relais de Transmissions Exceptionelles (France)
ASW	Anti-Submarine Warfare
AUTODIN	Automatic Digital Network
AUTOVON	Automatic Voice Network
AWAC	Airborne Warning and Control
BAOR	British Army on the Rhine
BICES	Battlefield Intelligence Collection and Exploitation System
BMEWS	Ballistic Missile Early Warning System
CCIS	Command and Control Information System
CDS	Chief of Defence Staff
CENTAG	Central Army Group
'Chop'	Transfer of NATO member's forces to operational command of SACEUR

CINCEUR	Commander in Chief Europe
CINCHAN	Commander in Chief Channel
CJCS	Chairman of the Joint Chiefs of Staff
CODA	Centre d'Opérations de la Défense Aériennes (France)
COEC	Command Operations and Exercise Committee
COFAS	Centre d'Opérations des Forces Aériennes Stratégiques (France)
COTAM	Commandement du Transport Aérien Militaire (France)
DCN	Defence Communications Network
DCS	Defense Communications System
DDN	Digital Defence Network
DEB	Digital European Backbone
DEFCON	Defence Condition
DOD	Department of Defense (US)
DPC	Defence Planning Committee
DSN	Defence Switched Network
DSCS	Defense Satellite Communications System
DSSS	Defence Secure Speech System (UK)
DSP	Defense Support Program (satellites)
EAM	Emergency Action Message
EBAU	Extended Battlefield Area Use
EBS	Emergency Broadcast System
ECCCS	European Command and Control Console System
EDG	European Deterrent Group
EMP	Electromagnetic Pulse
ELF	Extremely Low Frequency (radio)
ERCS	Emergency Rocket Communications System
ETS	European Telephone System
FATAC	Force Aérienne Tactique (France)
FSL	Field Storage Locations
GEP	Ground Entry Point
GOP	Generated Operations Plan
GPG	General Political Guidelines
GPS	Global Positioning System
GWEN	Ground Wave Emergency Network
HF	High Frequency (radio)
IDCSP	Initial Defence Communication Satellite Program
IFF	Identification Friend or Foe
IMS	International Military Staff
IONDS	Integrated Operational NUDET Detection System

IVSN	Initial Voice Switched Network
JCEC	Joint Communications and Electronics Committee
JCS	Joint Chiefs of Staff
JSTARS	Joint Surveillance Target Attack Radar System
JSTPS	Joint Strategic Target Planning Staff
LAN	Local Area Network
LDU	Limited Defensive Use
LERTCON	Alert Condition
LF	Low Frequency (radio)
LNO	Limited Nuclear Option
MAO	Major Attack Option
MC	Military Council
MCC	Missile Control Centre
MEECN	Minimum Essential Emergency Communications Network
MLF	Multilateral Nuclear Force
MNC	Major NATO Command
MOD	Ministry of Defence (UK)
MSC	Major Subordinate Command
NAC	North Atlantic Council
NACISC	NATO Communications and Information Systems Committee
NADGE	NATO Air Defence Ground Environment
NAEWF	NATO Airborne Early Warning Force
NARRS	North Atlantic Radio Relay System
NATO	North Atlantic Treaty Organisation
NCA	National Command Authority
NCCDPC	NATO Command and Control and Data Processing Committee
NCEB	NATO Communications and Electronics Board
NCP	Nuclear Capability Plan
NDAC	Nuclear Defence Affairs Committee
NEACP	National Emergency Airborne Command Post
NICS	NATO Integrated Communications System
NICSMA	NICS Management Agency
NJCEC	NATO Joint Communications and Electronics Committee
NMCC	National Military Command Centre
NMCS	National Military Command System
NOP	Nuclear Operations Plan
NORAD	North American Air Defence

NORTHAG	Northern Army Group
NPG	Nuclear Planning Group
NSC	National Security Council
NSTDB	National Strategic Target Database
NSTL	National Strategic Target List
NSVN	NATO Secure Voice Network
NUDET	Nuclear Detonation Detection System
NWAS	Nuclear Weapon Accounting System
NWS	Nuclear Weapon Storage Site
OTH-B	Over-The-Horizon Backscatter (radar)
PACCS	Post Attack Command and Control System
PAL	Permissive Action Link
PNL	Prescribed Nuclear Load
PNS	Prescribed Nuclear Stockage
PPG	Provisional Political Guidelines
PSC	Principle Subordinate Command
PTT	Post, Telegraph and Telecommunications (commercial communications)
QRA	Quick Reaction Alert
RAMSES	Réseau Amont Maillé Stratégique et de Suivre (France)
RBAU	Restricted Battlefield Area Use
RITA	Réseau Intégré de Transmissions Automatiques (France)
RNO	Regional Nuclear Options
SAC	Strategic Air Command
SACEUR	Supreme Allied Commander Europe
SACLANT	Supreme Allied Commander Atlantic
SAO	Selective Attack Option
SASP	Special Ammunition Supply Points
SATCOM	Satellite Communications
SCARS	SACEUR Control Alerting and Reporting System
SDS	Survivable Defence System
SEP	Selective Employment Package
SHAPE	Supreme Headquarters Allied Powers Europe
SIOP	Single Integrated Operational Plan
SITCEN	Situation Centre
SLFCS	Survivable Low Frequency Communications System
SSBN	Nuclear-powered Ballistic Missile Submarine
SSIP	Sub System Integration Project
STRIDA	Système de Traitement et de Representation des Informations de Defénse Aériennes (France)

TACAMO	'Take Charge and Move Out' (US SSBN Airborne Command Post System)
TARE	Telegraph Automatic Relay Equipment
TNF	Theatre Nuclear Forces
TTM	Terrestrial Transmission Media
TWU	Theatre Wide Use
UHF	Ultra High Frequency (radio)
UKADGE	UK Air Defence Ground Environment
VHF	Very High Frequency (radio)
VLF	Very Low Frequency (radio)
WEU	Western European Union
WHIDDS	War Headquarters Information Display and Dissemination System
WIS	WWMCCS Information System
WWMCCS	World Wide Military Command and Control System

Introduction

This book is a study of nuclear command and control in NATO and the strategy of flexible response. It looks at the political, military and politico-military dimensions of NATO nuclear command and control from the introduction of nuclear weapons into the Western European theatre in 1952 until the *de facto* abandonment of the strategy of flexible response at NATO's Defence Planning Committee meeting in May 1990.

The book serves a number of different functions. It is firstly an historical overview of the development of NATO's nuclear command and control systems. The chronological organisation of the early chapters explores the *ad hoc* and evolutionary, rather than revolutionary, means by which the NATO nuclear command and control system was constructed. Secondly the study presents a series of case studies – of NATO, US, British and French systems – which provide insight both about how NATO's overall nuclear command and control architecture attempted to resolve the problems of multinational operations and how it sought to reconcile differences and tensions between US and European system requirements. The use of case studies give the study a European focus and emphasis which has been hitherto missing from related studies.

The book also serves as a single case study of a multinational organisation responding to challenge and change. In this respect the strategy of flexible response provides a revealing context because of the very particular and exacting demands the strategy placed upon NATO's nuclear forces. However, it goes further to offer an analysis of the strengths and weaknesses of NATO's nuclear command and control systems with particular reference to the operational plans implicit in the flexible response strategy.

Through the examination of political and military dimensions of nuclear command and control the book provides a framework for consideration of thematic command and control issues both with respect to national forces and to the forces of a multinational military alliance.

Finally, the book offers some observations about the future of nuclear command and control in NATO in a greatly changed context. It does not attempt a descriptive update of NATO's nuclear command

and control system following the changes in the European theatre in the late 1980s (not least because programme inertia and a lack of clear strategic direction in NATO have left many of these systems largely unchanged and issues unresolved) but rather identifies the key issues raised by the study of this aspect of NATO's history and asks what relevance and lessons the experience of the past might hold for the future. In this respect, despite the changes, NATO faces consistent problems of reconciling national differences and US-European differences, of reshaping NATO's command and control system to meet a new strategy and doctrine, and of addressing persistent thematic problems in nuclear command and control including system interoperability, security and safety, and ensuring that command and control systems can underwrite the operations implicit in nuclear strategy.

In order to begin this analysis we first need to define and explain the central elements of the study, namely the strategy of flexible response and nuclear command and control. On December 14th 1967 NATO formally adopted the military strategy set out in Military Council directive 14/3 which became known as flexible response.[1] Although the actual text of Military Council directive 14/3 remains classified NATO documentation articulates the strategy as follows:

> The basis of this concept is that NATO should be able to deter, and if necessary, to counter military aggression of varying scales in any region of the NATO area; this can be secured only through a wide range of forces equipped with a well-balanced mixture of conventional, theatre nuclear and strategic nuclear weapons. The purpose of this balance of forces, while retaining the principle of forward defence, is to permit a flexible range of responses combining two main capabilities: to meet any aggression by direct defence at a level judged appropriate to defeat the attack, and to be prepared to escalate the level deliberately, maintaining firm political control, if defence at the level first selected is not effective. An aggressor must be convinced of NATO's readiness to use nuclear weapons if necessary, but he must be uncertain regarding the timing or the circumstances in which they would be used.[2]

Since 1952 the nations which commit nuclear forces to NATO have deployed a range of nuclear weapons with all three elements – air force, army and navy – of their armed forces designed to meet the operational requirements of flexible response.

In 1985 at probably the highest point of diversity in the nuclear arsenals these included 2700 free-fall bombs; 100 Ground Launched

Cruise Missiles [GLCMS]; 1870 artillery shells (8" and 155mm); 1072 Lance missiles, 198 Honest John missiles; 686 Nike-Hercules missiles; 220 Pershing IA and 54 Pershing II missiles; 332 depth bombs and 400 strategic Poseidon missile warheads dedicated to NATO, either deployed in the European theatre or stored outside Europe for use within the NATO theatre.

To these need to be added the 160 Polaris warheads, 406 gravity bombs and 120 depth bombs of the British nuclear arsenal and the French nuclear arsenal which comprised 18 land-based IRBMs, 176 submarine launched ballistic missile warheads, 200 gravity bombs and 120 Pluton missiles warheads. In total these amounted to 9,204 nuclear warheads either deployed in or committed to use within the European theatre by NATO nuclear forces.[3]

These weapons varied in explosive power from the sub-kiloton W-48 warhead used with the 155mm artillery to the B-28 free-fall bomb with a yield in excess of 1 megaton.[4] The delivery systems for these weapons varied in range from 14 kilometres for the 8" artillery shells to over 3,000 nautical miles for the Poseidon missiles.[5]

In terms of munitions NATO would seem to have possessed an array of nuclear weapons capable of responding to a range of levels of aggression and capable of selective escalation. However, the large number and diverse types of nuclear weapons deployed by NATO do not, of themselves, provide evidence of an ability to meet successfully the operational requirements of the strategy of flexible response. Of equal importance in any such assessment is the ability of NATO to command and control its nuclear weapons.

Defining command and control is not easy as the term is used in many ways and is itself subject to considerable variation.[6] In his study of the historical development of command and control Van Creveld prefers to use the term 'command',[7] whilst in many studies the functions of command and control are elaborated in the more usual terms 'command, control and communications' known by the acronym 'C^3',[8] and 'command, control, communications and intelligence' known by the acronym 'C^3I'.[9] A more recent term to have emerged is the simpler 'nuclear operations' intended to include the range of activities associated with command and control. In this thesis, given the broad definition set out below, the term 'command and control' and its acronym 'C^2' will be used as a common term for any and all of these variants.

The definition of command and control chosen is: an arrangement of facilities, personnel, procedures and means of information acquisition, processing, dissemination and decision-making used by national command authorities and military commanders in planning, directing and controlling military operations.[10] Since this study is concerned with NATO's nuclear forces the term is intended to apply specifically to nuclear operations, though it would also be proper to use the term for conventional operations.

Command and control is often understood by the simple analogy of military forces to human anatomy, where weapons represent the muscles of the body in that they are responsible for the fighting force, and command and control represent the brain, sense organs and nervous system.[11] Thus command and control has been characterised as having three distinct components:

> (a) command authorities, nuclear control systems and command centres to analyse data, make decisions, carry out directions and control forces; (b) sensors, including intelligence systems, providing inputs of warning and attack characterisation; and (c) communication links to distribute warning data and ensure the proper execution of commands.[12]

The importance of command and control in nuclear operations is difficult to overstate. If, to develop the human anatomy analogy a little further, shortcomings and failures of command and control are likened to failures of the brain, sense organs and nervous system the implications become apparent. Degradation of command and control may greatly impede the ability of a military force to carry out its operations, and, according to Blair, 'if command and control fails almost nothing else matters'.[13]

The analogy, while useful to illustrate the principles of command and control, should not be understood too literally. In NATO's command and control system there are many decision-making centres, multiple sensors and considerable redundancy of communications between principal command and force elements. The military system, in this respect, 'is far richer and more variegated than its analog'.[14] Nevertheless the hierarchical NATO command structures do lend considerable merit to the comparison.

The strategy of flexible response made very specific and exact demands on NATO's nuclear command and control system. To begin to understand these we first need a clearer statement of the strategy than that provided by NATO's public documentation. J. M. Legge in his 1983 study of NATO theatre forces for the American RAND Corporation provided just such a summary:

> . . . the flexible response strategy seeks to convince the Soviet Union that any level of aggression would be met not only with a direct response at that level, but that NATO would be prepared, if necessary, to raise the stakes rather than accept defeat. This demands not only maintaining a sufficient range of forces . . . but also convincing the Soviet leadership that NATO has the political will to use these forces in self-defence and, in particular, the preparedness, if necessary, to initiate certain quantum leaps in the level of conflict, the most crucial of which would be the crossing of the nuclear threshold. Finally, for the strategy to be credible it must not only be capable of conveying the desired message in peacetime so as to avoid a conflict occurring at all (pre-war deterrence), but also, should it fail . . . it must still offer the prospect of terminating the conflict at the lowest level consistent with NATO's objectives (intra-war deterrence).[15]

Enabling NATO to select nuclear response options considered appropriate to differing levels of aggression and to escalate levels of nuclear use in response to both conventional and nuclear attack required a command and control system which imposed high-level political control on nuclear forces, provided accurate and timely attack warning and characterisation, enabled meaningful decision-making, provided assured communications to and from the forces in the field, and offered the prospect of terminating the conflict under control. It further required a command and control system robust enough to carry out these functions in crisis and conventional and nuclear war. Moreover, as we shall see, the speed of modern warfare and the constraints of NATO's commitment to forward defence placed important pressures of time upon some elements of command and control making timeliness a significant determinant of operational capability.

It is consequent upon this that significant shortcomings, vulnerabilities and failures of NATO's command and control system would have adversely affected NATO's ability to carry out the controlled, selective and graduated escalation of nuclear use required by the strategy of flexible response. The credibility of flexible response was vital for the credibility of NATO's deterrent posture and because of this the question of whether NATO's command and control system could support the strategy was obviously a critical issue. In their major study of nuclear weapons operations Carter, Steinbruner and Zracket support this view: 'Deterrence turns on the credibility of the command system's ability to realise in practice the threats implicit in weapons and strategy'.[16] It is the primary purpose of this book to examine the strengths and weaknesses of NATO's nuclear command and control systems during the period in which flexible response shaped military planning in order to assess whether a gap existed between the operational flexibility promised by the strategy and the nuclear command and control system's ability to underwrite that flexibility.

Although classified studies of NATO command and control have been carried out, there are none in the open literature which give specific and detailed consideration to the subject prior to the early 1980s.[17] During the 1980s the growth of command and control as an important subject area in the field of strategic studies began to redress this shortfall. The work of Ball,[18] Blair,[19] Bracken,[20] and Steinbruner[21] has, in this respect, been central to the understanding of command and control at the strategic level.

The command and control of nuclear forces at the theatre level has not attracted the same level of attention. While there is a considerable literature on the command and control of conventional forces, of which Blumenson and Stokesbury,[22] Coulam and Fischer,[23] Cushman,[24] Stares[25] and Allard[26] are amongst the most important, this has not been matched by studies of theatre nuclear command and control. Only a small number of open source studies of NATO nuclear command and control were carried out during the 1980s, the most important of which were those of Charles,[27] Kelleher,[28] and Ball.[29]

Bracken's discussion concentrated on the relationship between NATO's strategy and the nuclear force structure, a relationship which, he argued in 1983, would be 'the greatest determinant of what happens in a future European crisis'.[30] It was Bracken's thesis that NATO's command and control system was 'so unstable and accident-prone that national leaders would exercise little practical control over it in wartime'. He argued that NATO's declared intention to use nuclear weapons if necessary was incredible if the decision had to be taken rationally by national political authorities. The credibility of NATO's use of nuclear weapons and hence, in his opinion, the deterrent value of NATO's theatre nuclear forces, stemed from the proposition that the existing command and control system would in crisis have behaved in such a way as to devolve control of nuclear forces down the command chain and subject nuclear forces to ambiguity of command authority. Under hostile conditions such delegation and ambiguity might lead to the use of nuclear weapons without high-level political control. Conscious of the risk of escalation that would accompany any use of nuclear weapons, Bracken asked rhetorically: 'What other command mechanism could possibly be built to invoke a nuclear conflict that, for all practical purposes, is tantamount to a regional doomsday machine?'[31] Bracken's discussion focused on the difficulties of maintaining high-level political control and preventing nuclear escalation. It suggested that NATO's command and control system was probably unable to provide the degree of centralised political control required by flexible response.

Charles provided an analysis of nuclear planning in NATO which focused on the political and logistical difficulties NATO would have faced had it intended to use nuclear weapons first in a conflict with Warsaw Pact forces. His broad conclusion was that the dilemmas facing political authorities and the difficulties of obtaining meaningful consensus around the question of using nuclear weapons may have precluded the use of nuclear weapons as a joint political option.

By contrast, his detailed analysis of NATO's military planning and the procedures for the release of nuclear weapons suggested that in preparing nuclear weapons for use military commanders may have limited the options of political leaders and may, in fact, have made it difficult for them to avoid using nuclear weapons. For Charles there was a real danger that NATO's strategy of flexible response, in propagating for deterrence purposes the belief that nuclear war could be controlled and limited (thereby making the use of nuclear weapons more credible or at least less irrational), had led to the development of a military posture in which nuclear weapons could be used even if political leaders did not actually wish to use them. He stated:

> . . . in its desire to add to the credibility of its nuclear deterrent [by planning and preparing for nuclear use], NATO may have created a situation in which it will have difficulty refraining from nuclear use even if it wants to.[32]

In this respect the conclusions of his analysis closely echoed those of Bracken.

If NATO ever was involved in the use of nuclear weapons in Europe Charles was sceptical about controlling escalation. Indeed he argued that: '. . . the idea that the use of nuclear weapons can be limited, fine-tuned, and precisely controlled is probably a self-delusion – quite possibly a dangerous one'.[33] Kelleher's brief study focused on the political and military channels for releasing nuclear weapons and looked at 'bottom-up' and 'top-down' release procedures which respectively concerned requests for nuclear use from field units and orders from high-level authority. Kelleher highlighted important questions concerning the responsiveness of theatre nuclear forces to political direction and the cohesion of the NATO alliance during crisis and warfare. She noted in 1987 that:

> The present system of command in NATO, especially on the thickly settled central front, is clearly less than satisfactory. Should war come, most observers foresee chaos, even without TNF use.[34]

Despite improvements to NATO command and control during the 1980s and greater consultation and planning between NATO allies, it was Kelleher's assessment that 'the critical observer is left with the sense of unresolved problems',[35] at least one of which was an assessment, like those of the foregoing, that NATO's command and control system might not be able to control nuclear operations. She concluded:

> So long as theatre nuclear forces are physically present in central Europe and widely distributed amongst allied forces under C^3 arrangements vulnerable to disruption or attack, the risks of uncontrolled escalation remains.[36]

Ball provided an overview of NATO's command structure and command and control apparatus and discussed problems of system complexity and the lack of command and control interoperability between the armed forces of the national allies within NATO. In his preamble he set out what he thought had been missing from previous discussions of NATO's command and control:

> . . . such critical issues as the extent to which these theatre [command and control] systems are designed to support crisis management and to function across the spectrum from peacetime operations through various levels of conflict to the large-scale employment of tactical and theatre nuclear weapons, and whether in fact their inability to perform as planned would actually exacerbate the dynamics of the escalation process, have not been addressed.[37]

His own study went some way towards addressing these questions. He discussed the problems and dilemmas of releasing and using nuclear weapons and concluded:

> . . . the decision-making structure and command and control arrangements spawned by this alliance [NATO] are quite unsuited

to the demands of rapid and timely decision and implementation that would pertain on the European battlefield.[38]

Ball also considered the vulnerability of NATO's command and control system which he considers to be high, and identified 60 critical targets which he claimed 'if destroyed or incapacitated would completely nullify any capacity for controlled, informed or responsive military operations within the theatre'.[39] Overall his assessment cast doubt on the notions of selective nuclear use and escalation control which were required by the flexible response strategy.

At least three important areas of concern were evident from these four studies. Firstly, the coalition nature of the NATO alliance which necessitated political consultation on the question of using nuclear weapons, with few exceptions, presented NATO with potentially insurmountable problems in seeking to arrive at a joint decision to use nuclear weapons. This situation could, if anything, have been exacerbated further by the time pressures which the dynamics of the battlefield imposed on political decision-making.

Secondly, the processes set in train by the need to prepare militarily for the use of nuclear weapons during a period of crisis or war risked the escalation of conflict to the nuclear level without firm political control. Similarly, military preparations for the use of nuclear weapons risked limiting the choices of political leaders to the point where they may have had little option other than to sanction their use.

Finally, the system complexity and lack of interoperability of national and inter-service command and control systems. together with their vulnerability to a range of threats, potentially precluded options of selective, finely-tuned or graduated escalation of nuclear operations by NATO.

To some extent each of these studies was limited as an analysis of NATO nuclear command and control. All four omitted or gave little space to discussion of British and French nuclear command and control which would certainly have had an important role in NATO operations. None pulled the main issues together into a comprehensive whole. Ball, Bracken and Kelleher gave insufficient attention to the political dimension, while Charles provided insufficient detail about the command and control system itself. Moreover, many questions remain unanswered, not only about the nature of NATO's command and control system itself and its performance across the spectrum from peacetime to large scale tactical and theatre nuclear use (identified by Ball) but also about the political problems inherent in directing a multinational nuclear force and using nuclear weapons as an exact instrument of policy.

This is not to imply any criticism of these studies. Each provided immensely valuable insight into nuclear issues and each has in its own way been instrumental in defining the field as a legitimate area of study. Moreover each pointed out important gaps in our knowledge and important areas for future study.

Nevertheless, in order to develop this understanding it is important to provide a study of NATO nuclear command and control which does include consideration of British and French command and control and which does bring together and develop the key issues of theatre nuclear operations. One approach which permits this is to adopt as the central theme a discussion of the relationship between nuclear command and control and the strategy of flexible response. By considering, at both the political and military level, the extent to which NATO's nuclear command and control system could support nuclear operations across the spectrum of possible nuclear missions and in a variety of scenarios, the full dimensions of theatre nuclear command and control can be explored.

The central inquiry of this book, outlined above, is intended to allow just this analysis to be undertaken. The book describes the development of NATO's nuclear command and control system and attempts to assess how the system could have been expected to perform in crisis and war. The study has been organised into six chapters.

The first chapter describes the political arrangements for operational nuclear decision-making in NATO. The second chapter provides an overview of nuclear command and control in NATO. The next three chapters are nation-specific case studies which deal respectively with American [as far as these relate to Europe], British and French nuclear command and control systems. Although France remains a member of NATO its forces were not after 1966 part of NATO's integrated military command and France at no time subscribed to the strategy of flexible response. Consequently Chapter Five stands as a distinct case study which not only discusses French command and control but also explores the unique relationship between the command and control of French forces and those of NATO. The sixth chapter assesses the performance of NATO's command and control system in crisis[40] and the wartime stresses to which the system could have been subject had NATO ever found it necessary to implement the military operations required by flexible response. The chapter explores both the political and military dimensions of nuclear command and control as well as issues of the politico-military interface between the two.

The conclusion draws together the issues raised by the study and asks whether they hold lessons for the present and future as NATO faces a changed and in many ways uncertain future.

At this point it is appropriate to make a number of statements about how the book has been written and its limitations.

Firstly, details of nuclear weapons command and control are amongst the most carefully controlled of all military information and since the author has at no time sought, or been given, access to restricted or classified material, this book is inevitably circumscribed. Nevertheless, there is sufficient open-source information available to piece together a reliable picture of the political and military elements of NATO's nuclear command and control system. The bulk of this material has been obtained from sources such as American government and military publications, some of which were obtained under the US Freedom of Information Act; NATO documents and publications; documents and publications from individual European NATO member governments; technical books, monographs, journals and conference papers; political and military memoirs; and newspapers. A vital adjunct to this body of information has been primary source material in the form of interviews and correspondence with political, military, academic, and commercial personnel whose positions, duties and studies have provided them with a knowledge of the subject which, in some areas at least, exceeds that available in the open literature. The information from these sources was given to the author on the condition that it would be used in this study only in an unattributable form. The study avoids, however, making any statements based on these unattributable sources alone.

Secondly, the study presents an assessment of the performance of NATO command and control systems in crisis and discusses the ability of the system to function in a range of wartime scenarios. This presents certain problems. Assessment of the behaviour of NATO's command and control system in crisis is considerably more difficult and less reliable than describing NATO's command and control system itself. There is some empirical evidence of the behaviour of NATO's command and control system during past crises which may be helpful, with certain caveats, in any assessment. Carter, Steinbruner and Zracket nevertheless urge caution in applying the lessons of past crises, and note that the nuclear command and control system has:

> . . . never been brought truly close to the edge of war . . . the behaviour of the command system near the brink of war, when the prospect of combat operations seems very real and even likely, is untested and unknown.[41]

There is, however, a considerable body of literature which presents empirical evidence of crises and decision-making in crisis.[42] This can be particularly instructive in assessing the decision-making of political élites during crises and as such may have some lessons for the political elements of NATO's command and control system. This literature contains much of value in assessing decision-making at the political-military interface, but has rather less to say about military decision-making during crises. To support this part of the study eight crises have been chosen from which to present evidence with some bearing on the present discussion. These are: the Hungarian Crisis of 1956; the Berlin Crisis of 1961; the Cuban Missile Crisis of 1962; the Arab-Israeli conflict of 1967; the Czechoslovakian Crisis of 1968; the Yom Kippur Middle East Crisis of 1973; the Cyprus Crisis of 1974; and, the Polish Crisis of 1980.

Some assistance may also be obtained from what is known of planning in NATO (for example the alerting process) and what can be gleaned from observation of wargames. Both, however, are unreliable surrogates for experience.

The issue of command and control performance in wartime is even more problematic. Since the NATO alliance was formed in 1949 it has not had to fight a conventional or nuclear war in Europe and therefore there is no direct empirical evidence upon which to base a discussion. Such a discussion is therefore difficult and fraught with uncertainty. Previous studies which have attempted comparable assessments of nuclear command and control systems in war have done so by describing the systems and inferring their behaviour during (where relevant) conventional and nuclear war from what is known of operational planning and adversarial capabilities.[43] These studies have of necessity been speculative and have avoided scenario dependence by considering a range of potential circumstances and assessing command and control capabilities in each. In the present study, when empirical evidence ceases to be available or reliable, the same approach has been adopted.

There are at least two further general limitations to studies of command and control. The first of these is the lack of relevant theory. Bracken has complained that 'the absence of even a rudimentary theory creates barriers to our understanding of command and control'.[44] Coulam and Fischer have similarly argued:

> . . . existing theories of organisation are not rich enough to provide precise assertions concerning the relationship between command and control activities and the quality of organisational performance.[45]

Foster provides a useful summary of the current state of theoretical research in command and control, but is still only able to conclude: 'If we agree on nothing else at this point than to agree on the need for a theory of command and control we will have made signal progress. Today there is no such theory.'[46]

The second general limitation is the inherent complexity of command and control systems and the consequent difficulty of analysis. Carter, Steinbruner and Zracket argue that this leads to intellectual segmentation in which knowledge of the system is fragmented amongst individuals. They note: 'even the most advanced experts and the most experienced practitioners are narrowly and incompletely informed. No-one understands the whole'.[47]

This does not mean that the study of command and control is necessarily unsatisfactory. Because open source material about the command and control systems and their capabilities is known, because weapons' effects are known or can be calculated, because planning gives clues as to how operations are intended to be conducted, and because a range of scenarios can be considered, analysis of the behaviour of command and control systems in crisis and war can – with due caution – be valid.

Finally, several notes are in order about specific limitations of this study. Firstly, unique problems are encountered when attempting to analyse a coalition of sovereign states as diverse as the NATO alliance. The military forces of eight member nations – Belgium, Greece, Italy, the Netherlands, Turkey, the United Kingdom, the United States and West Germany – had a primary nuclear role in NATO during the period under discussion. That is they were involved in training for, and may have been called upon to use US nuclear weapons under nuclear-sharing arrangements (throughout the book these forces are termed the bilateral nuclear forces). Assessing the command and control of such a diverse nuclear force necessitates inevitable simplification. Within the pages of a single book it is possible only to make generic assessments of most individual member nations. To redress this somewhat the analysis of US nuclear weapons earmarked for use by the British under bilateral nuclear-sharing arrangements explores the issue of bilateral command

and control in more detail. Moreover, the case studies of US and French nuclear command and control in the NATO theatre explore additional national issues and consequently widen the discussion.

Secondly, the subject of intelligence, which is often closely associated with command and control in the acronym C^3I, is important in considering the command and control of nuclear forces in NATO. For example, an aspect of intelligence such as strategic warning – that is warning of adversarial preparations for attack – was essential for the preparedness and efficacy of NATO's military operations.

Intelligence is, however, a subject complex and important enough to itself merit detailed study and this book has insufficient space for such analysis, other than to include rudimentary discussion of intelligence where it directly effects nuclear command and control in NATO.[48]

Finally, the relationship between nuclear command and control within the NATO theatre and the command and control of nuclear forces outside NATO needs to be mentioned. Events taking place within the NATO theatre would undoubtedly have influenced events in the command and control systems of forces outside the NATO theatre. It is simply not possible, given limitations of space, to discuss the global dimensions and interactions of this coupling[49] and this study does not attempt the task.

1 Political Control of Nuclear Weapons in NATO

INTRODUCTION

The uniquely destructive potential of nuclear weapons has necessitated that their use be governed only by the highest political authority. Each of the nuclear powers has consequently vested the authority to use nuclear weapons solely in the hands of their senior political leadership.[1] Because of this the role of political leaders in nuclear command and control assumes central importance.

In the NATO alliance two nations, the United States of America and the United Kingdom, historically provided nuclear weapons for the integrated military command. Ultimate authority over their use rested respectively with the American President and the British Prime Minister. Membership of NATO, of itself, did not in any way diminish the absolute authority of the two powers over their respective nuclear forces. Nevertheless, operational nuclear control in NATO was complicated by the multinational nature of the alliance and by the complexity of arrangements for sharing nuclear forces. The nature of nuclear control in NATO since the 1950s reflected the desire for alliance agreement on the use of nuclear weapons, given the gravity of such a step for all alliance members. American and European concerns about nuclear weapons in Europe, which have sometimes been in accord and sometimes not, have shaped the arrangements for nuclear control which ceded a measure of control to non-nuclear NATO members under certain circumstances.

To understand the role of political decision-making in the command and control of nuclear forces in NATO it is necessary to explore these arrangements in some detail. This chapter provides an historical consideration of the development of nuclear sharing and the collective political control of the use of nuclear weapons in NATO.

To do this the chapter is organised into four sections. The first considers the development of nuclear sharing in NATO since the introduction of theatre nuclear weapons (TNF)[2] into Europe in 1952. The three remaining sections address three related but distinct facets of the political control of nuclear weapons in NATO. The first of these is the arrangements for political consultation between NATO member nations on the circumstances and nature of the use of nuclear weapons. The second is the role of NATO members in the formulation of political guidelines for the use of nuclear weapons in Europe. The third is the organisation of NATO for crisis decision-making and the role of the member nations within that organisation.

NUCLEAR SHARING IN NATO

Less than a year after the signing of the North Atlantic Treaty in April 1949 which brought into being the NATO alliance, US President Truman issued a directive to his Secretaries of Defense and State to re-examine American security objectives in peace and war. This process led to the submission in April 1950 of NSC-68, a document which set out the nature of the Soviet threat to world peace and suggested that this could best be countered by a policy of containment which resisted Soviet expansion wherever it occurred. The new policy was to be underpinned by increased US military preparedness in the non-Soviet world.[3]

The preponderance of standing Soviet conventional forces in post-war Europe was perceived by the alliance as a threat to its security. The decision to forward-base US nuclear weapons in Europe was taken in recognition of the inability of the NATO alliance to provide an assured conventional defence against potential Soviet aggression.

Several factors underlined NATO's early reliance on nuclear weapons. Firstly, there was a growing feeling within the alliance that conventional force goals set at the North Atlantic Council (NAC) meeting in Lisbon in February 1952 – already revised downwards from 100 to 50 divisions – were unrealistic and unlikely to be met.[4] Secondly, research in the United States in the late 1940s and early 1950s had led to the development of small tactical nuclear weapons whose specifications were apparently well suited to battlefield use. Thirdly, military studies had provided strong arguments that theatre nuclear weapons could be used by an inferior armed force to defeat conventionally superior forces.[5]

By October 1953 the US National Security Council (NSC) authorised the US Joint Chiefs of Staff (JCS) to henceforth base their plans on using theatre and strategic nuclear weapons against conventional attack wherever this was deemed militarily advantageous.[6]

It is unclear exactly when the first US theatre nuclear weapons were introduced into Europe. It is usually accepted that this occurred in 1953, but there is evidence from General Laurus Norstad (Supreme Allied Commander Europe 1956–63) that: 'In fact the first nuclear elements were intoduced into the alliance in the summer of 1952.'[7]

The first short-range nuclear weapons to arrive in Europe were 280mm nuclear artillery.[8] Within a few years these had been joined by Honest John and Corporal short-range missiles (1953–5); atomic demolition munitions, the so-called ADMs, or nuclear mines (1953–5); Matador longer-range missiles (1954); intermediate-range Thor and Jupiter missiles (1958–9); and the USAF F-100 C/D (1955) and USN A-6 nuclear-capable tactical aircraft (1956) earmarked for European use.[9]

When US President Eisenhower took office in 1953 his 'New Look' security policies stressed a reliance on nuclear weapons as a means to justify substantial reductions in relatively costly conventional forces. NSC 162/2, the basic planning document of the new policy issued in October 1953, set out American aims as being to meet the Soviet threat to US security while at the same time avoiding 'seriously weakening the US economy or undermining [our] fundamental values and institutions'.[10] Theatre nuclear weapons provided the means to largely accomplish this in Western Europe.

In 1954, in response to NSC 162/2, NATO, under the direction of SACEUR General Alfred M. Gruenther, conducted a study to assess the implications of changing NATO's strategy to a greater reliance on nuclear weapons. The findings of this study, leaked to the press in July 1954, supported the change by noting:

1) warfare in the future would inevitably be nuclear;
2) the first atomic targets would be armed forces and military installations rather than major centres of population;
3) the peak of destruction would come at the outset of war; and,
4) therefore the outcome would be determined by the active forces in being.[11]

By the end of 1954 the North Atlantic Council (NAC) had authorised NATO commanders to plan to use nuclear weapons against the forces of the Warsaw Pact whether the latter used nuclear weapons or not. In a speech given at the Royal United Services Institute in London in late

1954, Field Marshal Bernard Montgomery, then Deputy SACEUR, made NATO's nuclear emphasis explicit:

> I want to make it absolutely clear that we at SHAPE [Supreme Headquarters Allied Powers Europe] are basing all our planning on using atomic and thermonuclear weapons in our own defence. With us it is no longer: 'They may be used', it is very definitely: 'They will be used if we are attacked.'[12]

Within three years NATO had formally adopted a Military Council strategy document known as MC 14/2 which committed NATO to the massive use of nuclear weapons in response to Soviet aggression.

Taken together, the steadily rising numbers of nuclear weapons in the American stockpile in Europe, the shift in strategy to a reliance on the early use of nuclear weapons, and the explicit nuclear plans of NATO's military forces – and hence an almost inevitably nuclear dimension to any conflict in Europe – led to a growing disquiet amongst many of the European NATO allies.

According to one author the policy of reliance on American theatre nuclear weapons for the defence of Western Europe was acceptable to the European allies provided:

> . . . that it represented an underpinning of the primary deterrent threat of the use of US strategic nuclear weapons against the Soviet homeland; that the weapons would be available for use against massed Soviet forces before they reached NATO territory; and that there was no prospect of a response in kind by Soviet forces.[13]

As the Soviet nuclear arsenal began to grow in the early 1950s these assumptions were placed in doubt. European anxiety was deepened by the 'Carte Blanche' war game which was conducted in 1955 with the purpose of reassuring the West European public of NATO's determination to maintain the integrity of NATO's territory in war. The exercise envisaged the use of nuclear weapons in the defence of the NATO alliance including their use on West German territory as a result of Soviet forces crossing the inter-German border. The 'outcome' of the exercise postulated some 1.5 million, mostly West German, civilian dead and a further 3.5 million injured.[14] The publication of these figures stirred widespread public and political anxiety about NATO's nuclear strategy and undermined the European willingness to allow the Americans to continue to dominate NATO's nuclear decision-making.

Many European allies, particularly the West Germans – who were likely to bear the brunt of any conflict – and the Benelux countries, were keen to seek a greater voice in both determining NATO's nuclear strategy and in controlling the use of nuclear weapons in Europe.

For the United Kingdom, which had detonated its first atomic weapon in October 1952, these concerns were also important. The United Kingdom, however, was already party to an informal bilateral 'gentleman's agreement' with the USA concluded in 1952 which promised consultation between parties prior to the use of nuclear weapons by the USA, and after 1952 by the UK.[15]

In France the continued hegemony of the USA in Europe and particularly the American nuclear dominance prompted calls for greater European participation in collective decision-making but also underlined the urgency of the development of a French nuclear weapon.[16]

On the American side, awareness of European fears – and American concerns about the prospects of further nuclear proliferation which European anxiety might engender – led to a willingness to accommodate European demands for greater control over nuclear forces in Europe.

At the December 1957 meeting of the NAC, held for the first time at the level of heads of government, the first steps were taken towards the sharing of nuclear forces in NATO and a greater participation of the European allies in decision-making. The decision was taken to stockpile American nuclear weapons in Europe for use by the European allies. The final communiqué of the meeting stated:

> . . . NATO has decided to establish stocks of nuclear warheads, which will be readily available for the defence of the alliance in case of need. In view of present Soviet policies in the field of new weapons, the Council has decided that intermediate-range ballistic missiles will have to be put at the disposal of the Supreme Allied Commander Europe [SACEUR]. The deployment of these stocks and missiles and arrangements for their use will accordingly be decided in conformity with NATO's defence plans and in agreement with the states directly concerned.[17]

The NAC called for the creation of a bilateral nuclear force in which the sharing of nuclear weapons was to be arranged on the following lines: the USA would provide nuclear warheads (and thereby retain its direct control over US nuclear munitions) whilst the particular

European ally participating in the arrangement would provide and operate the delivery system (aircraft, missile launcher, and so forth). This arrangement would naturally be extended on a bilateral basis between the US and any of the NATO allies willing to participate.

In order for the bilateral force to be created two separate bilateral agreements had to be concluded with each NATO country on whose territory US nuclear weapons were deployed or operated to provide the legal authorisation for American participation.

The first of these was a formal Program of Co-operation between the US and the respective European ally which referred specifically to one or more categories of nuclear weapons.[18] The agreement drew on Section 144b of the 1958 amendment to the US Atomic Energy Act (and became known as the 144b agreement)[19] which states:

> The President may authorize the Department of Defense, with the assistance of the [Atomic Energy] Commission to co-operate with another nation or with a regional defense organisation to which the United States is a party, and to communicate to that nation or organization such Restricted Data (including design information) as is necessary to –
>
> (1) the development of defense plans;
> (2) the training of personnel in the employment of and defense against atomic weapons and other military applications of atomic energy;
> (3) the evaluation of the capabilities of potential enemies in the employment of atomic weapons and other military applications of atomic energy; and,
> (4) the development of compatible delivery systems for atomic weapons.[20]

In effect each Program of Co-operation required US possession of nuclear weapons even though the weapons were dedicated to allied delivery systems, but permitted the European allies sufficient knowledge to train and operate the weapons and delivery systems as required.

The second agreement was a military service-to-service 'stockpile' agreement which set out the practical arrangements for the storage, security, control and operation of the weapons.[21]

Between 1957 and 1962 the US negotiated bilateral agreements with Belgium, Canada, France, Greece, Italy, the Netherlands, Turkey, the

United Kingdom, and West Germany (France ended the arrangement when it withdrew from NATO integrated military command in 1966).[22]

The debate about the 'arrangements' for the use of nuclear weapons of which the 1957 stockpile communiqué had spoken went to the heart of European anxieties about the American domination of nuclear decision-making. The 'arrangements' established by the bilateral agreements took two forms. Firstly, there was the separation of delivery system and nuclear warhead which enabled the USA to retain custody of nuclear warheads almost until the moment of use. An example might be a nuclear-capable artillery gun manned by European allied soldiers which only became a nuclear weapons system when the US handed over nuclear artillery shells.

Secondly there were systems – such as the Thor missiles deployed in the UK – in which both delivery system (missile) and weapon (warhead) were integral. In such a situation joint control was effected by a 'dual key arrangement' in which an officer from the United States and one from the relevant allied European power each held one of two keys necessary to operate the weapon. Without the co-operation of both – the argument went – the weapon could not function.

There were, at least in European eyes, a number of drawbacks to these arrangements. Firstly, there were large numbers of US nuclear weapons deployed across Europe over which the European allies exercised no control. Secondly, the procedures established for nuclear sharing in the bilateral agreements gave effective control to the USA over all nuclear weapons systems in Europe (bar a few in the UK's national stockpile) since the US could simply refuse to hand over warheads or use its 'keys' for dual key systems. Thirdly, by codifying US control the arrangements did little to discourage independent national nuclear efforts by the European allies. Finally, the meaning of 'control' itself was open to broader interpretation than that of simple custody and use implied by the bilateral agreements.

In the debates that followed the 1957 stockpile communiqué the nature of 'control' over the use of nuclear weapons in NATO was explored more fully. Osgood, in his analysis of the debates about sharing control in NATO, described a number of components of control which included:

> . . . manufacture of weapons; ownership of weapons; custody of weapons; manning and operation of weapons; military command of weapons; planning the strategy for using nuclear weapons (their

> number, type, and deployment and the contingencies, methods and targets of their employment); the political decision to use or not to use nuclear weapons; [and] the political decisions to govern the use of nuclear weapons in combat.[23]

Questions of the manufacture, ownership and custody of nuclear weapons revisited the proliferation debates which were taking place following the success of the UK in detonating its first nuclear device in 1952 and the French intention to follow a similar course. The US – and some in Europe – were concerned that a multitude of nuclear powers in NATO would lack political cohesion and be difficult to organise around collective defence.[24]

The manning and operation of nuclear weapons was, within the limitations discussed, addressed by the bilateral agreements and gave the European allies involved a measure of control in certain circumstances.

The military command of nuclear weapons and the military aspects of planning for their use will be discussed in the following chapter which considers nuclear command and control in NATO.

The keenest areas of interest for European political leaders were in obtaining a far greater voice in the planning of strategy for the use of nuclear weapons in NATO and in the political decision-making which governed the decision to use the weapons and how the weapons would be used. The way in which the USA responded to this demand and the subsequent evolution of collective political operational decision-making is the subject of the remaining sections of this chapter.

However, before developing this discussion it is necessary to bring the analysis of the arrangements for sharing nuclear forces up to date. The bilateral agreements of 1957–62 continue as the basis for present sharing arrangements to the present time (although there are no longer any 'dual key' systems deployed). Developments in strategy, the removal or introduction of nuclear weapons and other variables have necessitated the renegotiation of each agreement as required, but this has usually been done with the minimum of discussion and involved only marginal changes.[25]

Although the bilateral agreements determined the nature of nuclear sharing in NATO they did not temper the effective US control of nuclear weapons in NATO. While developments in the military command and political aspects of control were taking place (see below) there were also a number of initiatives to attempt to change the nature of nuclear sharing itself within NATO. Although these initiatives came to nought they are worth exploring since they throw light on European anxieties about the US dominance of nuclear decision-making.

The first to merit attention was known as the Gallois plan, named for General Pierre Gallois, an influential French military adviser and writer. Gallois suggested an arrangement in which the US would make nuclear weapons available to allied governments under some form of dual control with the firm agreement that the US would hand over her 'key' to the weapons under predetermined conditions of crisis or international tension.[26]

This would have the virtue of allowing the US to retain custody of the weapons in peacetime whilst ensuring – through automatic release – that nuclear weapons were at the disposal of individual European powers prior to war. Moreover, giving each of the European allies involved the power to retaliate with nuclear weapons, Gallois argued, would 'enhance the credibility of the threat of retaliation . . . for the nation whose vital interests were directly threatened to have the power to retaliate'.[27] Criticism of the Gallois idea, however, centred on the fact that the 'automatic' release of nuclear control to the allies could not be achieved in practice and that effective control would remain in US hands. Moreover, the enhanced deterrence promised by Gallois was by no means assured for reasons of only marginal interest here.[28]

Two more important initiatives were those designed to establish a European strategic nuclear deterrent. The first of these proposed a European Deterrent Group (EDG). This envisaged a European strategic nuclear force which would be jointly owned and controlled by a number of European allies (intended to include Belgium, France, Italy, Luxembourg, the Netherlands and West Germany). This force would be separate from both the strategic nuclear forces of the USA and the UK and controlled collectively by its European owners as a single military unit.[29]

The arguments in favour of the EDG included enhanced deterrence and the possibility of freeing the European NATO allies from US nuclear dominance. The arguments against were more numerous. Firstly, there were doubts about where the weapons for the force would come from. Would the US be willing (or able) to sell or supply nuclear weapons for the EDG over which it would subsequently lose control? Or would the UK (or France) be willing to place the fruits of their national efforts to develop nuclear weapons under collective control? Neither seemed likely.

Secondly, there were questions about command and control of the force and whether its use could be collectively agreed. Thirdly, the creation of a 'nuclear club' within NATO which did not include all the European allies threatened the political cohesion of the alliance. Finally, developing a European nuclear deterrent outside US control threatened the very commitment of the US to Europe upon which the alliance had been premised.

A second approach, in a similar vein, was the proposed creation of a NATO nuclear force. This entailed the establishment of a separate NATO nuclear force under NATO command and subject to collective rather than national release authority.[30] The proponents of a NATO nuclear force argued that the US should place some of its forces (strategic and tactical) deployed in the European theatre under collective control, together with weapons from the UK and France. The advantage of a NATO nuclear force was that all of the European allies would be involved in the decision-making as would the USA. This it was argued would:

> extend . . . the military entanglement of the US and Europe so as to enhance the cohesion of the alliance and insure it against Russian attempts to isolate Europe militarily or politically. Thus through an integrated nuclear force nuclear sharing could become an instrument of allied consolidation rather than disintegration.[31]

In effect this idea placed fifteen (the number of NATO allies at the time) fingers on the nuclear trigger and fifteen on the nuclear safety catch. A key question – and one which was central to the downfall of the idea – was whether the arrangement undermined deterrence by placing nuclear retaliation in doubt because of the time delay and command and control problems inherent in reaching collective agreement on the use of the force.[32]

The final proposition, and the one which came closest to full implementation, was the creation of a multilateral nuclear force (MLF). The proposal for the MLF was first articulated at the NATO ministerial meeting in Paris in December 1960.[33]

The MLF concept was to be developed in two phases. The first was to assign a number of American strategic nuclear submarines (and those of the UK when ready) to SACEUR who would be given the authority to launch the weapons under certain circumstances. The second was the creation of a multinational nuclear force which would operate a number of strategic nuclear weapons. This multinational nuclear force was ultimately intended to comprise a fleet of 25 merchant-type surface ships each of which would carry eight Polaris A-3 nuclear missiles.[34]

During the early part of the 1960s the MLF concept gained credence. Some American and, after 1962, all British ballistic missile submarines were assigned to NATO (a legacy which persists) and the multinational nuclear force project took shape. In October 1963 an MLF working group was formed consisting of representatives from Belgium, Greece, Italy, the Netherlands, Turkey, the UK, the USA and West Germany. By October 1964 the first of the MLF vesssels, the *USS Claude V. Ricketts* (the renamed *USS Biddle*) was launched as a test bed for the MLF concept. It was crewed by British, Dutch, Greek, Italian, Turkish, US and West German officers. The *USS Claude V. Ricketts* never did, however, become operational.

The assignment of some strategic nuclear submarines to SACEUR strengthened the US strategic guarantee in European eyes but the MLF concept with its implicit US control of the weapons (the missiles were in US custody on the vessel and could not be used without US authority) and its cumbersome command and control arrangements did nothing to address European desires for greater control.

The project was allowed to stall after 1964 by which time developments at the political level had begun to address European concerns about joint control of nuclear forces in NATO. Writing in 1964 Alistair Buchan predicted that:

> . . . the MLF will in the end prove to have been nothing but an expensive and time consuming detour on the road to a more effective system of political and strategic planning amongst the Western allies.[35]

He was right.

The more important advances in the joint control of nuclear weapons in NATO were to be made at the political level and it is to this that we now turn.

POLITICAL CONSULTATION

The basis for consultation in the Alliance lies in Article 4 of the North Atlantic Treaty which states:

> The parties will consult together whenever, in the opinion of any of them, the territorial integrity, political independence or security of any of the parties is threatened.[36]

The nature of consultation within the alliance has largely been shaped by the recommendations of a number of sequential (though not directly related) NATO reports and declarations. The development of what in 1970 the US Ambassador to NATO Harland Cleveland termed a 'doctrine of consultation'[37] has its origins in a three-man committee which assembled in 1951 primarily to consider the question of burden-sharing in NATO.[38]

The report compiled by the committee noted:

> The achievement of a closer degree of co-ordination of the foreign policies of the members of the North Atlantic Treaty, through the development of the "habit of consultation" on matters of common concern, would greatly strengthen the solidarity of the North Atlantic Community and increase the individual and collective capacity of its members to serve the peaceful purposes for which NATO was established.

It continued:

> Special attention must be paid, as explicitly recognised in Article 4 of the Treaty, to matters of urgent and immediate importance to the members of NATO, and to 'emergency' situations where it may be necessary to consult closely on national lines of conduct affecting the interests of members of NATO as a whole.[39]

Drawing on the work produced by the 1951 committee a subsequent committee, again involving three senior individuals (who became known as the 'three wise men'), met in 1956 explicitly to consider the role of consultation in fostering political co-operation within NATO.[40] Their report, named the Report of the Committee of Three on Non-Military Co-operation in NATO, was published on 13th December 1956. The Committee of Three attempted to define NATO consultation:

> Consultation within an alliance means more than exchange of information, though that is necessary. It means more than letting the NATO council know about national decisions that have already been taken; or trying to enlist support for those decisions. It means the discussion of problems collectively, in the early stages of policy formation, and before national positions become fixed. At best this

> will result in collective decisions on matters of common interest affecting the Alliance. At the least, it will ensure that no action is taken by one member without a knowledge of the views of the others.[41]

The report accepted, however, that there might be circumstances in which consultations would be impractical:

> . . . a member government should not, without adequate advance consultation, adopt firm policies or make major political pronouncements on matters which significantly affect the Alliance or any of its members, unless circumstances make such prior consultation obviously and demonstrably impossible.[42]

Two further documents throw light on NATO's consultative ethos, but each – with respect to the question of consultation – may be said to be little more than an affirmation of the foregoing and an attempt to reconcile practicalities with the desire for an ideal conception of the consultative process.

The Harmel Report, more formally known as '*The Future Tasks of the Alliance: Report of the Council*', was published on 14th December 1967. It considered a wide range of future tasks facing the Alliance and on the question of consultation it noted:

> As sovereign states the Allies are not obliged to subordinate their policies to collective decision. The Alliance affords an effective forum and clearing house for the exchange of information and views; thus, each Ally can decide its policy in the light of the close knowledge of the problems and objectives of the others. To this end the practice of frank and timely consultation needs to be deepened and improved.[43]

In 1974 the NATO heads of government agreed the Declaration of Atlantic Relations, a document which was intended to redefine the principles of transatlantic co-operation and to reorientate American interests towards Western Europe after almost a decade of preoccupation with the Vietnam War.

The declaration addressed a broad range of questions and on the subject of consultations stated:

> The Allies are convinced that the fulfilment of their common aims requires the maintenance of close consultation, co-operation and mutual trust, thus fostering the conditions necessary for defence . . . In the spirit of friendship, equality and solidarity which characterises

> their relationship, they are firmly resolved to keep each other fully informed and to strengthen the practice of frank and timely consultation by all means which may be appropriate on matters relating to their common interests as members of the Alliance, bearing in mind that these interests can be affected by events in other areas of the world.[44]

While each of the four documents affirm the desirability of consultation each is equally aware of the necessary limitations of consultation. None of the documents is obligatory in nature nor is any specific as to just what type of decisions call for consultation. This reflects the nature of NATO itself which, as a voluntary alliance, has no authority or means to compel compliance of any member with any of its agreements or declarations on questions of consultation as on other issues. One author has usefully noted that:

> NATO is not a supranational organisation with real clout over its members. NATO does not have mandatory authority. It can advise, exhort, or exercise moral suasion, but it cannot direct its members to implement even those measures they have already agreed to undertake.[45]

There are likely to be further limits to NATO consultation which have their basis in the unavoidable precedence of the national priorities of individual member nations. It has been noted, for example, that:

> . . . the political consultation process [in NATO] is always restrained somewhat by the natural reticence of governments about disclosing too much, even to closest allies, through fear of a 'leak', a desire to keep national options open as long as possible, or an unwillingness to expose domestic policy-making to the pressures of outside viewpoints.[46]

Another study which considered operational decision-making in practice added that:

> . . . timing and secrecy can hinder consultation, and member governments, particularly when faced with urgent decisions vital to themselves, may feel that they can do no more than inform their allies shortly before, or even during, action.[47]

These reservations and limitations to the consultative process should not, however, obscure the fact that consultation on a wide range of matters became an established and well-practised process in the NATO alliance.

A study by Kirgus which looked at both the legal basis and historical practice of consultation in the Alliance concluded the following:

> NATO practice establishes that there is an obligatory norm for consultation within the alliance before a member government makes a final decision on a course of action that (1) could have a direct, adverse effect on the alliance's collective ability to defend against an attack from outside the treaty area, or (2) could significantly enhance the military capability of those states presumed most likely to attack.[48]

According to the same study this 'obligatory norm' extends to:

> . . . such NATO consultations as are feasible when an ally contemplates action within or outside the treaty area that could lead, in the short run, to armed confrontation with the Soviet Union or could seriously exacerbate an existing confrontation with it.[49]

While these clearly identify areas in which NATO would be expected to be in political agreement, it nonetheless provides evidence that consultation, at least on these issues, was the rule rather than the exception.

As the report of the 'Committee of Three' made explicit in 1956 there was, however, an equally important question concerning the nature of consultation and whether this process could be said to be a collective discussion before national positions become fixed or an information-exchanging (or providing) process in which notification is given of decisions already taken. The evidence from the Kirgus study discussed above is that the nature of consultation in NATO – even in those areas of decision-making where consultation has been usual – has tended towards the latter. This has been the case even when the decisions taken establish a course of action that is for practical purposes irreversible.[50]

Against this background of the consultative ethos of NATO the question of consultation on the role and use of nuclear weapons in Europe represents a special case in which the alliance members were manifestly unequal and in which the nuclear powers undoubtedly dominated decision-making.

The reasons behind the nuclear-sharing accommodation of the Americans which led to the creation of NATO's multinational nuclear force have been discussed above. It is equally important to understand the ways in which NATO has sought to provide a means for European allied input into policy and operational nuclear matters. Specifically, this means the ways in which NATO consultation on nuclear matters has been organised.

The first clear expression of the willingness of the US and UK to consult with their non-nuclear allies on questions governing the role and use of nuclear weapons occurred at the Ministerial-level meeting of the North Atlantic Council (NAC) held in Athens from 4th – 6th May 1962. The Americans and British agreed to continue to provide nuclear weapons to the alliance in sufficient numbers to meet the threat to alliance security. In addition they agreed to consult with their allies, time and circumstance permitting, before initiating the use of nuclear weapons.[51]

To provide practical substance to such an undertaking the NAC resolved at the Athens meeting to 'set up special procedures which will enable all members of the Alliance to exchange information concerning the role of nuclear weapons in NATO defence'.[52]

A number of initiatives followed which were, at least ostensibly, greatly to increase the European role in nuclear decision-making in NATO. At the May meeting of the NAC in 1963 the Council approved the assignment of a European deputy to SACEUR for nuclear affairs.[53] The deputy was to be supported by a staff drawn from all the NATO European member countries. The significance of this move lay in institutionalising European involvement in the military side of nuclear decision-making which up to that point had been almost entirely the province of SACEUR, an office held only by American officers.

A second initiative designed to broaden European military participation in nuclear matters was to assign European officers to liaise with the Office of the Joint Strategic Targeting Planning Staff (JSTPS) based at Offut Air Force Base in the USA, the headquarters of Strategic Air Command (SAC).[54] The JSTPS was tasked with allocating targets to all US strategic forces, including those committed to NATO. The assignment of European officers to liaise on the European elements of targeting provided an important European input into the nature of nuclear use in Europe (targeting in NATO is discussed in more detail in Chapter 2).

It was in the political arena, however, that more important initiatives to involve European members in nuclear decision-making were to be taken. At a special meeting of Defence Ministers in May 1965 the US Secretary of Defense Robert McNamara proposed that a select committee of alliance members should be set up at Ministerial level to improve collective participation in nuclear decision-making and to develop a more effective consultative process.[55]

McNamara initially proposed limiting the participation to five members but this suggestion was unpopular with the Europeans. As the result of a compromise a special committee of all the defence ministers who wished to participate was agreed. This was eventually to include ten ministers. This temporary committee was subsequently divided into three smaller groups which were respectively assigned the tasks of looking at crisis management (including exchanges of data), communications and nuclear planning in NATO.

According to the then US Ambassador to NATO further negotiations then narrowed participation in the nuclear planning group 'which everyone knew was destined to be the center ring of this little circus' to five members. These were the defence ministers of the US, UK, West Germany, Italy and one other. The fifth member was decided by the drawing of lots, after which the Turkish defence minister took the seat.[56]

In four meetings between February 1966 and September 1966 the Nuclear Planning Group of the NATO Special Committee of Defence Ministers (known informally as the McNamara Committee) discussed a broad range of issues including the role of nuclear weapons and the difficulties of the nuclear planning process. Unprecedented American candour undoubtedly contributed to the success of these meetings and the emergence of a consensus around the idea of institutionalising a forum for discussing nuclear planning in NATO.[57]

In December 1966 the Defence Planning Committee created a two-tier forum to meet the McNamara Committee recommendation. The first tier of the forum was called the Nuclear Defence Affairs Committee (NDAC), a plenary body open to all interested allies. The NDAC would receive reports from the second tier seven-member Nuclear Planning Group (NPG). From the outset it was envisaged that the NPG would be the forum in which the detailed discussions of nuclear matters would take place.[58]

The reason for the two-tier structure was the need to reconcile the demand for most of the European allies to have a voice in nuclear policy and planning with the American (and particularly McNamara's) preference for a relatively limited forum to facilitate effective and rapid decision-making.

All of the NATO members except France, Iceland and Luxembourg joined the NDAC. The composition of the NPG was initially decided in the following manner. Permanent participation was granted to the

US, UK, West Germany and Italy (France having chosen by this time to opt out of NATO's integrated military structure altogether). Three additional places were to be occupied by the remaining allies on a rotational basis of 18 months' duration.[59]

Opposition from some European quarters to the composition of the NPG was accommodated with a provision for review of the arrangements after 1 January 1970.

The first meeting of the NPG was held in Washington on 6th – 7th April 1967. Since then the NPG has continued to meet twice yearly at ministerial level.

The NPG was the principal architect of the refinement and elaboration of the arrangements for nuclear consultation in NATO which took place over the following years. Part of the work of the NPG concerned the development of political guidelines for the use of nuclear weapons and this is discussed in the following section of this chapter. The NPG simultaneously considered the question of the consultative process itself and sought to clarify the commitment of the 1962 'Athens guidelines' to consult with allies, time and circumstances permitting.

At meetings of the NPG in The Hague in April 1968 and in London in May 1969 discussions considered, *inter alia*, under what circumstances consultation on the issue of nuclear use would take place, who would participate, and what limitations would be applied to the consultative process.

The resultant Provisional Political Guidelines (PPGs) remained classified but some of their recommendations have been published in US Congressional testimony. The guidelines were apparently explicit that the normal forum for consultation on the question of the use of nuclear weapons would be the Defence Planning Committee (though it is equally possible that it could have been the NAC given the relationship between the two committees). The guidelines also provided a procedural prescription to be followed should a request for the use of nuclear weapons be forthcoming either from a member government or a Major NATO Commander (MNC):

> . . . any request for the use of nuclear weapons in defence of NATO either from a member government or from a Major NATO Commander, and any possibility for the use of nuclear weapons in defence of NATO by a nuclear power, would be communicated immediately to the NATO governments and to the Defence Planning Committee . . . where member governments would be able to express their views, in particular on the political and military objectives of

> the proposed use of nuclear weapons, the methods of use and the possible consequences either of use or non-use. These views would then be communicated to the nuclear powers concerned, and the decision of the nuclear powers would be conveyed to the allied governments, the North Atlantic Council and the Major NATO Commanders.[60]

Moreover, the guidelines reaffirmed the commitment of the US and UK to consult with allies before using nuclear weapons, with the necessary caveat of 'time and circumstances permitting'. Explicit recognition was given to the importance of the views of those NATO allies most directly involved in any decision to use nuclear weapons. The guidelines were thought to provide that:

> . . . special weight be given to the views of the NATO country most directly affected – that is, the country or countries on, or from, whose territory nuclear weapons would be employed; the country or countries providing the nuclear warheads; and the country or countries providing or manning the contemplated means of delivery.[61]

The foregoing provided a comprehensive consultative process, yet at the same time recognised limitations which could override the desire to consult, particularly on issues as grave as the decision to use nuclear weapons.

There were grounds for scepticism about the machinery of consultation in NATO as it pertains to operational decision-making. The distinction between policy decision-making and operational decision-making is, in this respect, a very important one. NATO's consultative fora have proved to be generally effective for peacetime decision-making on matters of policy and planning. Indeed the organisation and procedures of NATO's bureaucracy which have developed over the years have evolved primarily to meet the needs of peacetime decision-making.

At ministerial level both the North Atlantic Council and the Defence Planning Committee met but twice a year. These meetings were highly formalised and followed an agenda set well in advance. One study described ministerial NAC and DPC meetings thus:

> Ministers generally present set-piece speeches, routinely endorse a number of reports, and approve a communiqué largely drafted in advance of the meeting . . . Important and sensitive subjects are usually addressed in bilateral discussions and in executive sessions

> where attendance is restricted and discussions not encumbered by formalised ministerial agenda.[62]

Both the NAC and the DPC were formal and stylised fora and it was unlikely that either could in practice be adapted for swift and hard-headed decision-making. Both could serve as a low-level discussion forum obviating the need for multiple bilateral contacts, but sensitive or urgent discussions – such as those about the use of nuclear weapons – could probably not have been carried out in either.[63]

POLITICAL GUIDELINES

Rudimentary guidelines concerning the use of nuclear weapons in Europe were first issued by NATO in 1957.[64] These were circulated to member governments at the time when the first non-American NATO allies were being trained to use nuclear weapons as a result of the Eisenhower stockpile arrangements which had been formally agreed by the NAC in December 1957.

Very little is publicly known about these guidelines except that they stressed the use of nuclear weapons in response to Soviet aggression. It is apparent, however, from the debates which subsequently took place and from the guidelines which were subsequently issued by NATO, that few hard questions concerning the nature, timing and circumstances of the use of nuclear weapons were addressed by the 1957 guidelines.

The origins of the political guidelines which subsequently addressed the question of nuclear use in NATO can most usefully be traced back to 1962. The series of bilateral agreements between the USA and European allies between 1957 and 1962 on the sharing of nuclear weapons formed the basis for an alliance agreement on the question of consultation. At the 1962 Athens ministerial session of the NAC pertinent sections of the various bilateral agreements were codified in what became known as the 'Athens Guidelines'. The NATO communiqué from the meeting stated:

> So that all member states may play their full part in consultation on nuclear defence policy, it has been decided to set up special procedures which will enable all members of the alliance to exchange information concerning the role of nuclear weapons in NATO defence.[65]

The 'Athens Guidelines' thus provided the basis for subsequent consultation within NATO on nuclear weapons questions. The guidelines also provided the continued assurance that both the USA and UK would make nuclear weapons available to the alliance in the numbers and types required to meet the needs of NATO's defence.[66]

The formal adoption by NATO of the strategy of flexible response in 1967 was a significant change from the trip-wire strategy articulated in the military document MC 14/2 to which NATO had previously adhered. The 'Athens Guidelines' predated the adoption of flexible response and thus had been premised, almost entirely, on the massive and prompt use of nuclear weapons in response to Soviet aggression. The new strategy implied a far greater need for political control over the employment of nuclear weapons and thus asked more of the consultative mechanisms of NATO. Within NATO there was as yet, however, no detailed planning or doctrine to support the new strategy.

At the NPG meeting of 6th – 7th April 1967 the NATO defence ministers agreed to '. . . initiate a number of studies to help clarify important questions related to the use of tactical nuclear weapons'.[67]

The studies were to look in detail at the various possible scenarios of the use of tactical nuclear weapons in the NATO theatre. By April 1968 at the third meeting of the NPG (the second having taken place at Ankara in September 1967) the US Secretary of Defense Clark M. Clifford proposed that on the basis of the studies thus far completed the NPG should proceed to develop political guidelines for the use of tactical nuclear weapons.[68] The NPG communiqué noted:

> The Ministers reviewed a number of studies in the field of nuclear planning conducted since their last meeting, with special reference to the use of various types of tactical nuclear weapons in the defence of the North Atlantic Treaty area, and welcomed the progress that has been made. They gave directions for the elaboration of further guidelines for the NATO military authorities in respect of the tactical use of nuclear weapons.[69]

The work was to be completed in stages. The initial work was to address the question of the initial use of tactical nuclear weapons by NATO.[70] At a later date the work would be expanded to include consideration of the follow-on use of nuclear weapons.

The work on the initial use of nuclear weapons by NATO was broken down into several separate studies to be undertaken by different NATO allies. The US addressed the question of the 'demonstration' use of nuclear weapons (that is the use of one or a very small number of nuclear weapons for the sole purpose of signalling political resolve); the West Germans looked at battlefield use; the Italians considered the use of atomic demolition munitions (ADMs); and the British assessed the maritime use of tactical nuclear weapons.[71]

By the next meeting of the NPG on 10th – 11th October 1968 there had been sufficient progress made by the respective groups for the defence ministers to propose drawing the main findings together into a 'single document that would ultimately provide authoritative political guidance on NATO initial use'.[72]

The British and West German defence ministers Denis Healey and Gerhard Schroeder jointly presented this composite document (which became known as the Healey-Schroeder Report) to the meeting of the NPG on 28th-30th May 1969. The clearest statement of the contents of the report – which remains classified – was made by J. M. Legge in his 1983 study of NATO theatre nuclear forces. According to Legge:

> The paper examined in detail the various options open to NATO: defensive use (ADMs or nuclear air defense); battlefield use; use in an 'extended geographical area' (in other words beyond the battlefield well into Warsaw Pact territory); maritime use; and so on.[73]

Although there were initial American reservations about parts of the paper – particularly the scale of initial use by NATO which the Europeans saw as very limited, but which the US thought should include larger-scale options – these were largely ironed out at a trilateral meeting between the US, UK and West Germany which occurred before the next meeting of the NPG in November 1969. Moreover, the comments of the other NPG members gave rise to more than 100 amendments to the initial draft of the Healy-Schroeder report. These reservations and amendments were taken into consideration before the production of the revised paper which was approved at the November NPG meeting. Known as the Provisional Political Guidelines for the Initial Defensive Tactical Use of Nuclear Weapons by NATO (shortened to the 'PPGs') the document was subsequently endorsed by the NDAC and DPC in December 1969.[74]

It is now known that the PPGs did not attempt to define the nuclear threshold, that is the point at which NATO would use nuclear weapons. Rather they were an attempt to think through the likely options for NATO nuclear use and the difficult questions NATO would have to face should the prospect of using nuclear weapons become real. Thus, according to one author, they '. . . constitute[d] an enhancement and explanation of the strategy of flexibility in response'.[75]

Legge suggests that the word 'provisional' was included in the title of the guidelines to indicate that NATO defence ministers considered the guidelines open to subsequent alteration or updating and that the second phase of the work – preparing guidelines for the follow-on use of nuclear weapons – was still to be carried out.

Work on the follow-on use of nuclear weapons began at the same NPG meeting in November 1969 which had endorsed the first phase guidelines. Ministers decided, as before, to undertake a series of studies to investigate the types of possible use and the geographic locations of such use of follow-on nuclear forces. Eight studies were initiated, each involving groups of three or four allied member nations. The findings of the studies were presented to the NPG between 25th – 26th May 1971 and the 15th – 16th May 1973. Once again Legge provides the most authoritative statement of their findings:

> . . . a single underlying message emerged with remarkable clarity from virtually every study. The studies indicated that follow-on use of TNW by NATO in the form of selective strikes against Warsaw Pact forces could result in short-term military advantage in the area concerned, and quite possibly a pause in the conflict; but if the Warsaw Pact responded with a nuclear attack on a similar (or greater) scale, neither side would gain a significant military advantage.[76]

The conclusion that the follow-on use of tactical nuclear forces might, if the Soviets responded in kind, lead to a military impasse, proved hard to accept for those in the Alliance who wished to see a continued reliance on the use of tactical nuclear weapons. There followed more than a year of 'difficult debate' within the Alliance before a second-phase report of the follow-on studies could be presented to the rest of the allies in July 1974.[77]

The conclusions of the follow-on use report echoed those of the PPGs in that:

> . . . follow-on use should have the same purpose as initial use (to persuade the enemy to cease his aggression and withdraw), and the nature of use should therefore still be selective and be designed to meet this political requirement.[78]

At the June 1975 meeting of the NPG the ministers gave directions for the 'conduct of the third and final phase of this [the political guidelines] program'.[79] The third phase was intended to draw together the PPGs, the work on follow-on forces, and the consultative guidelines (discussed in the previous section). The third phase was also to incorporate guidelines on the use of ADMs which had been the subject of a separate study after the adoption of the PPGs in November 1969, and work undertaken on theatre strike forces.[80]

The question of theatre strike forces referred to the longer-range nuclear missiles and aircraft systems deployed in the European theatre which had been excluded from consideration in the discussion of tactical nuclear forces. Prior to the adoption of flexible response theatre strike forces had been earmarked for use as an element of the large-scale use of nuclear weapons envisaged by MC 14/2. The adoption of flexible response raised questions about the uses of longer-range forces in the selective and gradually escalating use of nuclear weapons envisaged by the new strategy. The subject of the use of theatre strike forces was addressed in a series of US-West German studies before a paper entitled 'The Role of the Theatre Nuclear Strike Force in Allied Command Europe' was presented to the NPG on 29th – 30th October 1970.

The phase three program to consolidate the diverse guidelines and reports into a single authoritative document of political guidance for NATO on the use of nuclear weapons was delayed, however, by those who had reservations about one or more of the contributory documents. Amongst these reservations were concerns about the development of nuclear weapons technologies and their impact on the assumptions and findings of the various guidelines and reports.

Moreover, the scale of the American nuclear arsenal in Europe was itself subject to growing US Congressional scrutiny culminating in an amendment, proposed by Democratic Senator Sam Nunn, to the US Military Procurement Authorisation Act of 1976. The amendment required the US Secretary of Defense to report to Congress on the size and nature of the American nuclear stockpile in Europe and on the concepts which would govern the possible use of tactical nuclear weapons in the European theatre.

As a consequence of NATO's subsequent preoccupation with the size and nature of the nuclear arsenal necessary to support the strategy of flexible response, the development of detailed doctrine and planning, which had been the aim of much of the NPG's deliberations from 1967–77, was concluded and phase three of the guidelines programme postponed.

The subsequent history of the development of political guidelines in NATO is less clear and is likely to remain so until another insider, such as Legge, provides further details or until the relevant material is declassified. According to Daniel Charles, writing in 1987, the PPGs were supplanted by General Political Guidelines (GPGs) prepared by the NPG. Charles suggested, though his source was unclear, that:

> GPGs, as they are called, are more comprehensive than the PPGs and reflect many of the changes in NATO's nuclear stockpile during the 1970s, but do not overcome any of the fundamental problems associated with nuclear use.[81]

According to the 1989 *NATO Facts and Figures* the GPGs represented the completion of phase three consolidation in that they drew together the diverse guidelines and reports described in the foregoing.[82]

The PPGs were not intended to be operational orders, but instead were an attempt to think through the problems and issues associated with the use by NATO of nuclear weapons in order to illuminate the political decision-making process concerning the use of nuclear weapons in a crisis, during a period of conventional conflict or in response to a Soviet nuclear attack.[83]

The guidelines essentially addressed two types of questions. The first considered the political goals in any use of nuclear weapons, such as the avoidance of nuclear escalation, the cessation of hostilities, the withdrawal of Soviet forces, or the restoration of deterrence.

The guidelines also addressed questions of the nature, timing and circumstances of nuclear use – taking into account different weapons types and different geographic locations for both initial and follow-on use – necessary to meet the desired political objectives. The guidelines, inter alia, almost certainly prohibited the use of nuclear weapons within a certain distance of allied cities and limited such attacks in scale and yield. Thus the guidelines appeared to be both prescriptive and proscriptive.[84]

By these means the guidelines were also intended to address some of the inevitable political-military tensions involved in any contemplated use of nuclear weapons and thus to create the capacity to cope with these circumstances as they unfolded in a major crisis or war in Europe.

The nature of the NATO alliance, discussed above, makes it clear, however, that nothing about the guidelines was in any way binding on the member nations. The guidelines provided a framework, mutually arrived at by the member nations and therefore presumably largely acceptable to all, within which NATO might act in times of crisis and war.

To the extent that they were prescriptive there was perhaps an attendant risk, despite intentions, that the guidelines might be followed in a crisis or wartime situation rather more rigidly than circumstances dictated. In uncertain and confused circumstances the guidelines would at least have had the attraction of a high degree of allied support and agreement.

ORGANISATION FOR CRISIS DECISION-MAKING IN NATO

The Defence Planning Committee (DPC) was the principal forum for NATO consultation and co-ordination in periods of crisis. The procedures and organisation which supported the DPC in crisis deliberation evolved in response to the changing strategy of the alliance. Prior to the adoption of flexible response the trip-wire strategy of massive retaliation neither permitted nor required detailed and complex consultation between allies. The strategy of flexible response, however, envisaged controlled and escalating responses to aggression and therefore required:

> . . . considerable consultation during a period of rising tension and right through the full operational spectrum of response . . . [and] effective arrangements within the Alliance, in the broad field of crisis management, to provide all the information required as a basis for these consultations.[85]

The withdrawal of France from the integrated military structure of NATO in 1966 and the subsequent transfer of NATO headquarters, home of the North Atlantic Council and the Defence Planning Committee, from Paris to Brussels (and the Supreme Headquarters Allied Powers Europe [SHAPE] from Versailles to Mons) gave NATO the opportunity to consolidate all the various elements associated with defence planning and operations at a single political headquarters and a single military headquarters.[86]

The new NATO headquarters constructed on the site of an old military airfield at Evere outside Brussels contained offices and committee rooms for the North Atlantic Council and Defence Planning Committee and for the Military Committee which was moved from Washington DC (USA) to be co-located with the NAC. An underground situation centre, known as SITCEN, was also built below the new headquarters.[87]

The SITCEN was purpose-built to contain its own communications centre, computer processing, information display and high security meeting rooms where alliance leaders could meet in an emergency.[88] It was envisaged early on that in an emergency the DPC or NAC would meet for discussions in the consultation room of the SITCEN. Detailed briefings on political and military developments in a crisis would be given in the presentation room of the SITCEN.

It is from the SITCEN that direct, rapid, secure communications were established with national capitals and with the Major NATO Commanders (MNCs) SACEUR, SACLANT and CINCHAN.[89] The SITCEN was operated on a daily basis during peacetime by a multinational military and civilian staff which worked around the clock. Routine operations were under the control of the Military Committee.[90]

The arrangements and procedures to support the operation of the SITCEN in times of emergency were largely the responsibility of two committees both of which acted in consultation with high level military authorities. The first, known as the Command Operations and Exercise Committee (COEC) was:

> . . . responsible to the DPC for developing crisis management procedures and effecting command and control machinery – including arrangements for the exchange of information and intelligence data, devising alert procedures and for developing guidance on the actions which ought to be taken in times of crisis.[91]

The COEC was responsible for supervising the NATO crisis management machinery and for organising periodic exercises for testing the procedures and machinery (such as the biennial WINTEX series which ended in 1989).

The second committee, known as the NATO Joint Communications and Electronics Committee (JCEC) was the highest level communications policy body in NATO and was:

> . . . responsible to the DPC for ensuring the availability and effectiveness of communications systems necessary to permit members of the Alliance to consult – both in peace and during a crisis . . .[92]

The procedures for crisis decision-making set out by the COEC have been evolving since the adoption of the flexible response strategy.

These procedures essentially dealt with problems of information and intelligence sharing which were crucial to crisis consultation and decision-making.

One of the most serious problems was the need to exchange information and intelligence to ensure that each of the member nations involved in consultation was using a common data base of information about the prevailing circumstances. The nature of the procedures for disseminating information has been summarised as:

> . . . providing for intelligence coming from a capital to be transmitted to the NATO HQ in Brussels and the headquarters of the three Major NATO Commanders. Items of a particularly significant nature are disseminated immediately by capitals to all terminals on the NATO communications system.[93]

It would thus appear that the COEC specified pre-agreed crisis procedures for the sharing of information to all NATO member nations and the MNCs. While there was clearly an operational requirement and desirability for a common database it is nevertheless naive to believe that all member nations would share all of their information and intelligence equally from the early stages of a developing crisis. In the discussion of consultation above it was evident that national and domestic priorities play a major role in limiting the extent of information and intelligence sharing.

The COEC also devised procedures to try to provide for the political and military evaluation of shared information and intelligence, and, equally vitally, for the exchanging of such evaluations, particularly those made by individual NATO member governments. Exchanging evaluations was thought to provide 'a very necessary insight into the minds and attitudes of all member governments'.[94] It was also thought to be desirable for the DPC (or NAC) to arrive at a common interpretation of events and circumstances in order to facilitate agreement on appropriate actions.

An essential prerequisite for crisis decision-making in NATO was reliable, rapid and secure communications between each of the allied member nations, NATO headquarters and the Major NATO Commanders. When the NATO headquarters were moved to Brussels in early 1967 steps were taken to initiate a complete review of NATO's arrangements for crisis management and command and control.[95]

Up to that time the dedicated communications systems of NATO were constructed to reflect the needs of the strategy of MC 14/2. Since the trip-wire strategy required little in the way of consultation NATO constructed a network of point-to-point military communications links (ACE-HIGH) to ensure that the nuclear release orders could be transmitted to the commanders responsible for using nuclear weapons.[96] NATO headquarters before 1967 had no direct means to communicate with the heads of member nations and communications had to go through their delegated representatives.

The McNamara Committee which met for the first time in December 1965 to address questions of nuclear planning and policy also, as we have seen, addressed the question of the ability of NATO communications to facilitate consultation and the exchange of information and intelligence.

The report of the communications sub-group of the McNamara Committee recommended the establishement of a NATO-wide communications system to link NATO headquarters with the capitals of the member nations and the Major NATO Commanders.[97]

The programme which resulted from this initiative became known as the NATO-wide Communications System. It consisted of a full duplex (simultaneous send and receive facility) teleprinter system linking the SITCEN with each of the NATO member capitals and the MNCs. The system was controlled from a Marconi TARE (Telegraphic Automatic Relay Equipment) switching centre in the SITCEN and used Marconi Myriad 11 Computers to operate it.[98]

In order to avoid the obvious vulnerability of having a single switching centre in Brussels SITCEN to control the entire network a duplicate was constructed at the NATO headquarters at Northwood in the United Kingdom. The entire system was in operation by 1969.[99]

Although when first introduced the NATO-wide Communications System fulfilled a vital role and gave NATO a capability it did not previously possess its shortcomings were soon apparent. The TARE was a simple teletype message system which could provide neither voice nor data communications. Moreover, it was, according to at least one author, 'inflexible and not able to adequately meet NATO's command and control requirements in a crisis'.[100]

In 1969, coincidentally, the military SHAPE technical centre at the Hague presented to NATO a common-user communications network – known as ACENET – designed for Allied Command Europe. It was agreed that the ACENET concept provided the kind of communications capabilities required throughout the NATO theatre and consequently the ACENET concept was revised and expanded before being presented in 1970 as a proposed NATO Integrated Communications System known thereafter as NICS. In March 1971 the NICS concept was approved by the NAC and a management agency – known as the NICS Management Agency or NICSMA – was established to implement it.[101]

The requirements of the NICS programme were that it should 'provide rapid, secure, reliable, flexible and survivable communications for the quick exchange of information'.[102] The users were to include all the political, civil and military authorities supporting NATO to facilitate mutual consultation, crisis decision-making and the transmission of orders.

The NICS concept was based on a common user automatically switched, grid network involving a variety of transmission media to provide voice, telegraph and data traffic. Briefly explained, the system was based on providing a network of transmission systems, some new and some already in existence, to link switching and access nodes which in turn were linked to political, civil and military users. The purpose of the grid network system was to provide high system damage absorption and survivability by both the dispersal and multiplicity of the switching and access nodes and the redundancy of the numerous transmission systems between nodes. Still further survivability was then achieved by hardening the switching nodes and keeping nodes and user sites geographically separate. A communications pathway between any two (or more) subscribers could, in theory, be found by the computerised switching nodes even if large sections of the system had been damaged or destroyed during conflict.

The NICS was also intended to provide enhanced user facilities. For example, the system provided the usual telephone links between subscribers but in addition the system operated an automatic preferencing scheme in which users with high priority could, on dialing an engaged number, override the connected call if the latter involved lower priority subscribers. Moreover, a limited number of important subscribers were intended to have a variety of additional capabilities including secure voice, 'hot-line' type direct communications, conference calls and broadcast calls (in which one subscriber is able simultaneously to reach large numbers of others).

Similar features were intended for message transmissions though these were controlled by message distribution centres (MDCs). MDCs were also intended to provide a facility for the automatic distribution of messages to all addresses.[103]

The huge cost and scale of the NICS programme led to a decision to implement the system in two stages. Stage I was intended to provide the basic subsystems for the NICS and Stage II was intended to integrate the separate subsystems into a single system and to enhance the quality and survivability of the various system elements.

The history of the implementation of Stage I NICS, which was not fully operational when flexible response was abandoned, has been one of delay and specification compromise. As a result the system fell short of achieving the capabilities set out by the NICS concept in the early 1970s with consequent implications for crisis decision-making and command and control.

Stage I NICS, estimated to cost $500 million at 1972 prices,[104] was intended to comprise the following elements:

i) the Initial Voice Switched Network (IVSN), consisting of 24 operational access switches to provide an enhanced telephone system;
ii) the Telegraph Automatic Relay Equipment (TARE) comprising 18 operational TARE sites to provide improved message distribution;
iii) the NATO Satcom III subsystem consisting of three satellites, 21 static ground terminals and two transportable terminals;
iv) the Secure Voice Network to provide secure voice capability for about 1600 subscribers;
v) the Terrestrial Transmission Media making use of existing ground-based systems such as the ACE-HIGH troposcatter network (see Chapter 2), the civilian PTT network, the new (then) CIP-67 network which provided line-of-sight links in the central front region of NATO where there is a concentration of subscribers;
vi) the Subsystem Integration Project which provided the ancillary facilities needed at each site to ensure that all of the subsystems could subsequently be integrated.[105]

It was envisaged that Stage I NICS would be completed by 1980,[106] but this proved an optimistic forecast. The development of Stage I NICS in fact proceeded in the following way:

i) Initial Voice Switched Network

The contract for the IVSN was drawn up in 1973–4 and in 1976 was awarded to the American company ITT North Electric, with the first operational switches due in 1978.

Delays occurred which were due to NATO member nations failing to provide national IVSN sites on time and to software problems with the IVSN computers. In 1981 ITT North Electric announced that an initial examination of the software had led to the discovery of errors which would entail rewriting 100000 lines of computer code incurring a delay of four or five years. The errors had led to doubts about the reliability of the software which was prone to instability and frequent crashing.

For NATO the delay was unacceptably long and as a result a compromise was reached. Some of the technical specifications were reduced in return for a guaranteed completion date for all 24 switching nodes of the end of 1984. Completion of the IVSN did not finally occur until 1985. Moreover, to meet the deadline 'some of the originally specified overall capabilities and reliability levels . . . had to be reduced'.[107]

ii) Telegraph Automatic Relay Equipment

The TARE system of NICS needs to be distinguished from the TARE system which formed the basis for the NATO-wide Communications System which became operational at the end of the 1960s.

The contract for the NICS TARE system was drawn up in 1974 and given to Litton (USA) in 1976. The first TARE was delivered in 1978, but software problems subsequently plagued the development programme. The NICSMA, aware of the seriousness and scale of the software problems Litton were experiencing, recommended to NATO that the contract be terminated, but this request was denied.

NICSMA enforced the contractual testing of the TARE programme in the early 1980s but the system proved too unstable and was prone to crashing far too frequently to meet test specifications. In March 1983 NICSMA again tried to persuade NATO to terminate the contract and again failed. Subsequently NATO relaxed some of the specifications of the TARE in return for guaranteed completion of the first up-to-specification TARE by December 1983. The final TARE was not operational until 1988 and concerns about the system specifications remained.[108]

iii) NATO Satcom III

The NATO Satcom III programme was intended to place three NATO satellites in orbit and to expand the existing ground stations (serving NATO I and NATO II satellites) to 21 sites and in addition to procure two transportable ground terminals.

The contract was given to Ford Aerospace in 1973 and the three NATO satellites were successfully launched on time: NATO IIIA in April 1976; NATO IIIB in January 1977; and NATO IIIC in November 1978.

The upgrading of the 12 existing ground terminals and the addition of nine other stations was completed by 1983. By 1985 all 21 fixed stations were operational. The number of transportable terminals was reduced to one and that was operational by 1985.[109]

iv) NATO Secure Voice Network (NSVN)

The NSVN contract was given to Siemens and AEG Telefunken (FRG) and implemented in two stages. The first stage, completed by 1980, provided a dedicated manually-switched system to major user locations in NATO. By early 1988 the system had been automated and integrated into the IVSN providing secure voice communications between as many as 1600 subscribers.

v) Terrestrial Transmission Media (TTM)

The initial phase of the TTM programme, intended to enhance the capabilities of systems already available or under development in NATO, was completed by 1989. Seven projects to improve the capability of the ACE-HIGH troposcatter system were completed by early 1988. The CIP-67 [110] system, developed and deployed outside the aegis of the NICS programme, has also been completed.

vi) Subsystem Integration Project (SSIP)

As a composite project to oversee the provision of facilities to enable the integration of subsystem elements of the NICS programme the SSIP was still incomplete in 1990. The various subsystem integration facilities were not yet ready and thus the Stage I NICS remained formally incomplete.

Plans for Stage II NICS for the integration and enhancement of the subsystems developed and deployed in Stage I were under consideration by early 1977. In the autumn of 1978 thirteen volumes of systems architecture and plans, developed by NICSMA, were handed to the NJCEC and a year later they were approved. The overall cost of Stage II was put at $750 million (at 1979 prices) with the implementation due to be completed by 1995.[111]

In 1983, however, the NICS Stage II architecture was declared infeasible, and, as such, no longer a viable basis for the completion of the NICS programme.[112] The programmes both to enhance existing NICS capabilities and to integrate the subsystems into a single network were therefore placed in some doubt.

Technical staff who worked with the NICSMA office up to the summer of 1985 suggested however that considerable changes to the development of NICS were taking place. For the two years following the rejection of the Stage II architecture NICS enhancements were being based on a command and control improvement plan of the major NATO Commanders. During this period some enhancements to the capability and reliability of existing systems, such as the IVSN and TARE, were made but no progress was made towards the integration envisaged by Stage II. Simultaneously NICSMA was working on a new NICS architecture slated for completion by the year 2000.[113]

A major difficulty facing NICSMA in this task was the competition for funding within NATO with other command and control initiatives including the Command and Control Information System (CCIS) and the modernisation of the Air Command and Control System (ACCS).[114] In 1984, in an effort both to end the patchwork of committees dealing with command and control issues within NATO and to provide a single authority to oversee command and control procurement in NATO, a decision was made to integrate the various NATO committees with responsibilities for command and control into a single authority. The NATO Joint Communications and Electronics Committee (NJCEC) which dealt with communications but not information systems, the NATO Command and Control Data Processing Committee (NCCDPC) which dealt only with information systems, the NATO Communications Electronics Board (NCEB) which dealt with military interests in communications and electronics, and NICSMA were integrated.

The resulting committee, known as the NATO Communications and Information Systems Committee (NACISC), was approved by NATO in November 1985 and held its first meeting in May 1986.[115] The agency running the NICS programme was thus renamed the NACISA.[116]

The implementation of Stage II NICS, as a consequence of budgetary constraints and the new organisation of command and control management within NATO, developed along different lines from those proposed in the late 1970s.

The new architecture for NICS development involved a two-tier survivability concept. The first, designed to provide the necessary communications for periods of crisis and war, included utilising the systems then available to NATO and connecting these with national defence networks to provide additional NATO-wide communications and information systems survivability. These would be supported by constantly improving national PTT networks.

The second, designed to assure communications in nuclear war, involved a greatly reduced essential user network involving appropriate transmission systems such as mobile SATCOM terminals which have enhanced survivability against nuclear weapons effects and resistance to electronic warfare.[117] This system was not scheduled for completion before 1995 and therefore does not inform this discussion.[118]

CONCLUSIONS

This chapter has attempted to show why and how arrangements for nuclear sharing in NATO came about and, having done so, how the nuclear powers and NATO's other member nations have balanced their respective needs with respect to the control of nuclear weapons.

The resulting compromise, for such it is, met through complex arrangements for collectively determining nuclear policy, as well as through consultative arrangements for the operation of nuclear weapons in the European theatre, the European desire for a voice in the control of nuclear weapons deployed in Europe. On the other hand the nature of the NATO alliance, the arrangements for nuclear sharing and the limits of collective decision-making were such that the nuclear powers were effectively free to act autonomously if the deliberations of NATO in any way threatened to undermine or delay nuclear weapons operations deemed necessary by the nuclear powers.

This chapter has also considered the processes by which NATO would have attempted to reach collective agreement on the use of nuclear weapons and on the nature of that use. It has further discussed at some length the arrangements which were introduced by NATO to facilitate crisis decision-making through information sharing and to control NATO military (including particularly nuclear) forces through the NICS command and control system.

The discussion of the NICS system – developed by NATO to meet the theatre-wide needs of the political control of military operations –

serves as a small case study of how the implementation of a complex command and control architecture can be subject, during the transition from design to deployment, to the pressures of economics, national parochialism, and the need to reconcile design specificiations with achievable system capabilities. The result was a compromise system – still incomplete by 1990 – which provided NATO with less capability than was promised but considerably more than NATO had before the system was introduced.

The purpose of this chapter has been to place political decision-making at the heart of the consideration of the command and control of nuclear forces in NATO and to introduce the capabilities and limitations of the arrangements and systems for the political control of nuclear weapons.

2 Nuclear Command and Control in NATO

INTRODUCTION

This chapter discusses the organisation of nuclear forces in NATO. It develops the discussion of collective decision-making at the political level, which was explored in the previous chapter, by focusing on the military level and seeking to explain how the forces were organised for collective defence under the strategy of flexible response.

The chapter is organised into five sections. The first looks at the command structure of NATO and at the nuclear forces deployed within this structure. The second looks at the command and control systems which linked the various elements of the command structure. The third addresses the question of nuclear targeting to show how NATO reconciled the overlapping and competing targeting plans of its various nuclear forces. The fourth deals with the question of early warning and attack assessment to show how NATO as an organisation obtained and disseminated intelligence information and how this affected preparations for collective defence. The fifth deals with NATO nuclear alert procedures and describes NATO preparations for nuclear combat.

COMMAND STRUCTURE

Under the strategy of flexible response two NATO command structures existed side by side. The first was the national command structure of each of the individual member nations. The second was the formal collective NATO command structure to which national forces were nominally assigned in peacetime.

The parallel command structures existed, in principle, to assist the smooth transfer of military forces from national to NATO command in crisis or wartime. Since, for reasons of space, it is not practical to discuss each of the national command structures here (see Chapters Three and Four for an insight into US and British command respectively) this section concentrates on the NATO command structure.

In peacetime NATO's military forces were subordinate to the political North Atlantic Committee (NAC) and Defence Planning Commitee (DPC) through NATO's Military Committee (MC) shown in Figure 2.1.

The Military Committee was the 'highest military authority in the North Atlantic Treaty Organisation'[1] and comprised the Chiefs-of-Staff of each of the member nations except France (which nevertheless had a military mission attached to the committee). The Military Committee had the peacetime task of 'recommending those measures considered necessary for the common defence of the NATO area'.[2] It was served by an International Military Staff (IMS) which was made up of military and civilian personnel from each of the NATO member nations (except France) and acted as the executive arm of the Military Committee ensuring the implementation of policies and decisions made by the committee.

The NATO theatre itself was divided into three command areas: Allied Command Europe (ACE); Allied Command Atlantic (ACLANT); and Allied Command Channel (ACCHAN). Each of the three NATO commands was headed by a major NATO commander (MNC). ACE was commanded by the Supreme Allied Commander Europe (SACEUR), ACLANT by the Supreme Allied Commander Atlantic (SACLANT) and ACCHAN by the Commander-in-Chief Channel (CINCHAN).

ACE was NATO's principal military command. Its military forces operated over an area bounded by the northernmost part of Norway, the southernmost part of the Mediterranean, the westernmost part of the European Atlantic coast and the easternmost part of Turkey. SACEUR, as commander of ACE, thus had the dominant military role in NATO and, although his authority was theoretically subordinate to the Military Committee, in practice this was not so.

The office of SACEUR was always held by a senior US military commander whose stature and authority derived from the dominance of US military forces in Europe and the importance of the US commitment to Europe.[3]

One author has described the relationship of SACEUR and the Military Committee in the following terms:

> The Military Committee is overshadowed by SACEUR. No Chairman of the Military Committee has acquired the status of a

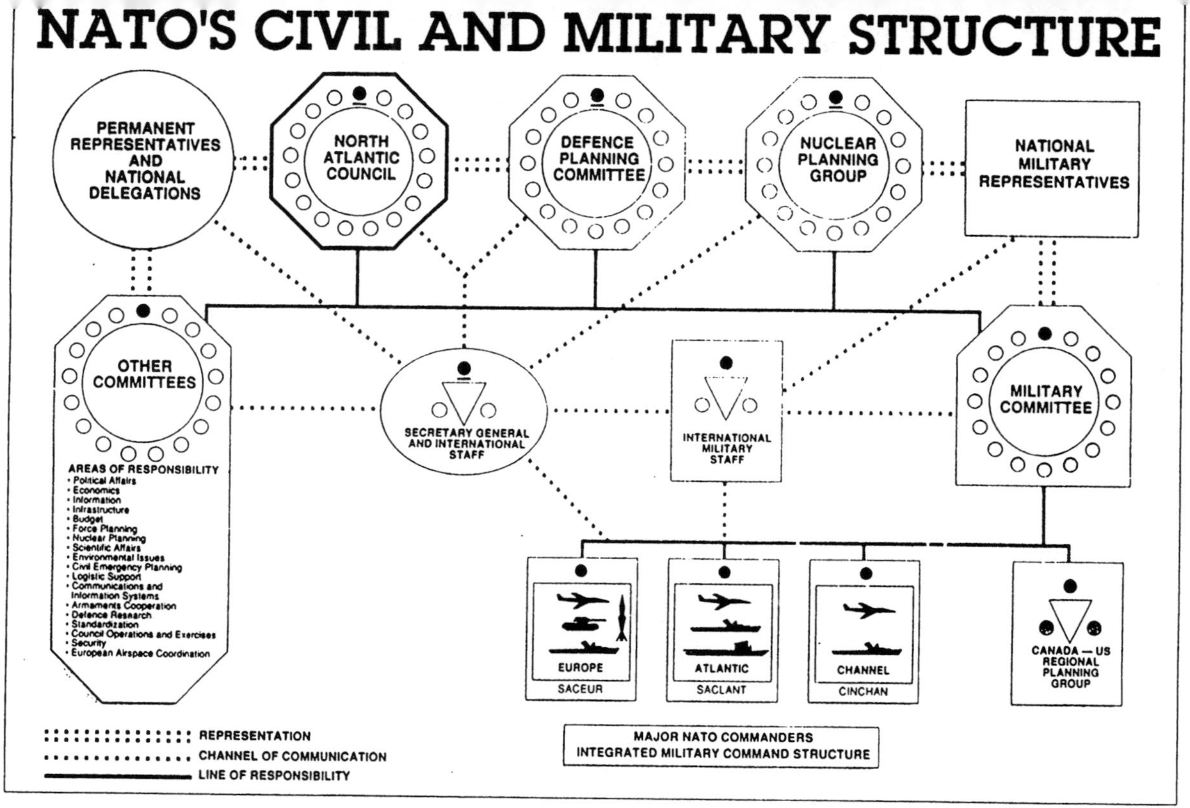

Figure 2.1 NATO Civil and Military Command Structure

Source: *NATO Facts and Figures* (NATO Information Service, 1989).

> SACEUR . . . It is to SACEUR, not the Military Committee, that the council (NAC) and allied governments look for military advice. SACEUR symbolises the US commitment to the defence of Europe and has the political and organisational clout to develop, negotiate and put into action NATO defence plans. SACEUR has the authority to deal directly with allied governments and with the [NATO] Secretary General, bypassing the Military Committee, and he exercises that right regularly.[4]

In practice the usual means by which NATO developed and co-ordinated defence planning was for SACEUR and allied national defence staffs to reach an agreement which could then be formally approved by the Military Council.[5]

SACEUR was headquartered at SHAPE, the Supreme Headquarters Allied Powers Europe, based at Casteau (Mons) in Belgium.[6] Figure 2.2 shows the military commands which were formally subordinate to SACEUR. In addition to SHAPE, Allied Command Europe comprised four major subordinate commands (MSCs) – AFNORTH, AFCENT, AFSOUTH and UKAIR – the ACE Mobile Force (Land) [AMF(L)], and exercised operational control of NATO's Airborne Early Warning (AEW) forces (see below).

AFNORTH, headquartered at Kolsaas (Norway), was the command area of Allied Force Northern Europe extending from the Finmark/North Cape region to the north, through Norway, Jutland and Schleswig-Holstein to the Elbe river in Northern Germany. It included the Baltic, Danish Straits and the seas adjacent to Norway and Denmark. AFNORTH had three Principal Subordinate Commands (PSCs): Allied Forces Southern Norway (SONOR) based at Oslo; Allied Forces Northern Norway (NONOR) based at Bodo; and, Allied Forces Baltic Approaches (BALTAP) based at Karup, Denmark.

Both Denmark and Norway had a policy of not allowing nuclear weapons or foreign bases on their soil and neither country was party to bilateral nuclear arrangements with the US. This did not, however, mean that AFNORTH was a non-nuclear command. The US operated naval and maritime air nuclear forces in the region and, to a limited degree, Denmark and Norway participated in the infrastructure to support these forces.[7]

AFCENT was the command area of the Allied Forces Central Europe which extended from the Elbe river in the north to the borders of Austria and Switzerland in the south. Although geographically the smallest of the three ACE regional commands AFCENT was the area

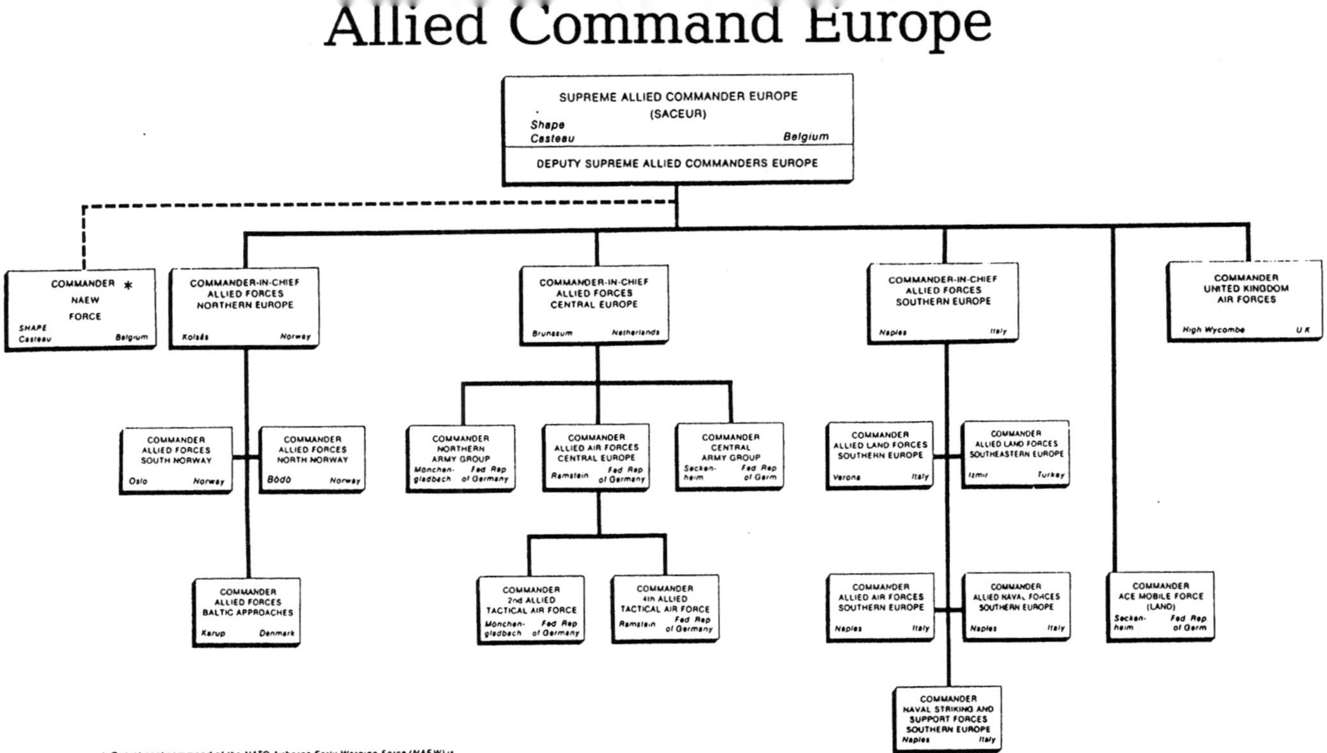

* Operational command of the NATO Airborne Early Warning Force (NAEW) is shared by SACLANT, CINCHAN and SACEUR. SACEUR acts as Executive Agent in promulgating operational command directives and is designated as Executive Agent in the day-to-day exercise of administrative control

Figure 2.2 Allied Command Europe (ACE) Command Structure

Source: *NATO Facts and Figures* (Brussels: NATO Information Services, 1989).

of the greatest concentration of military forces (conventional and nuclear). AFCENT had its headquarters in Brunssum in the Netherlands and had five Principal Subordinate Commands (PSCs): Allied Air Forces Central Europe (AAFCE) based at Ramstein; Northern Army Group (NORTHAG) and the Second Allied Tactical Air Force (2ATAF) both based at Moenchengladbach; and Central Army Group (CENTAG) and the Fourth Allied Tactical Air Force (4ATAF) based at Heidelberg, all in Western Germany. Operationally 2ATAF and 4ATAF were subordinate to AAFCE.

AFCENT commanded both US and British nuclear weapons, the former being provided for Belgian, British, Dutch, and West German forces as well as for use by the US.

AFSOUTH was the command area of Allied Forces Southern Europe which covered Italy, Greece, Turkey and the entire Mediterranean Sea. AFSOUTH was geographically the largest of the three ACE MSCs and had its headquarters at Naples in Italy. AFSOUTH has five PSCs: Allied Air Forces Southern Europe (AIRSOUTH) based at Naples; Allied Land Forces Southern Europe (LANDSOUTH) based at Verona in Italy; Allied Land Forces Southeastern Europe (LANDSOUTHEAST) based at Izmir in Turkey; Allied Naval Forces Southern Europe (NAVSOUTH) based on Nisida island in the Bay of Naples; and the Naval Striking and Support Forces Southern Europe (STRIKEFORSOUTH) also based at Naples.

AFSOUTH was, like AFCENT, a nuclear command in which nuclear forces were operated by the US in conjunction with Greece, Italy and Turkey.

UKAIR was a unique ACE MSC in that it was essentially a single service (RAF), single nation MSC and had no subordinate PSCs. UKAIR was based at High Wycombe in the UK and provided both air defence and attack airforces for employment within ACE. UKAIR operated British nuclear weapons.

The remaining two MSCs in ACE, AMF(L) and NAEWF, were non-nuclear commands and need not detain us here.[8]

In order to develop the discussion of the command structure more fully it is necessary to trace the command heirarchy below the level of PSC. To do this for all the NATO commands would be prohibitive of space so AFCENT has been chosen as an illustrative example.

Forces in the AFCENT command region were in the period of this study divided into a northern region commanded by NORTHAG (army) and 2ATAF (airforce) and a southern region commanded by CENTAG and 4ATAF.

NORTHAG comprised the following army corps: 1st Belgian Corps; 1st British Corps (BAOR); 1st West German Corps and 1st Netherlands Corps.[9] These were supported by 2ATAF aircraft drawn from the same national military forces.

CENTAG comprised the 2nd and 3rd West German Corps and the 5th and 7th US Army Corps, supported by 4ATAF aircraft drawn from the same national military forces.[10]

Army Corps were arranged along the AFCENT region in a 'layer-cake' formation facing what was the central front of the inter-German border (see Figure 2.3).

Defence in each of the corps segments was the responsibility of the corps deployed. Divisions, brigades and battalions at their command comprised national forces organised along national lines. At each level there were liaisons with air commands to co-ordinate air-land battle operations.[11]

The second of NATO's major NATO commands (MNCs) ACLANT commanded an area which extended from the North Pole to the Tropic of Cancer and from the coastal waters of North America to those of Europe and Africa, though not including the British Isles and the Channel. The office of SACLANT, as with SACEUR, was always held by an American officer and was headquartered at Norfolk in Virginia, USA. There were six Principal Subordinate Commands comprising ACLANT (see Figure 2.4).

The Western Atlantic Command (WESTLANT) based at Norfolk itself comprised the Submarine Force Western Atlantic Area (SUBWESTLANT) and the Ocean Sub-Area Command (OCEANLANT) both based at Norfolk; the Canadian Atlantic Sub Area Command (CANLANT) based at Halifax, Nova Scotia; and the Island Commands of Bermuda (ISCOMBERMUDA) and Greenland (ISCOMGREENLAND) based respectively at Hamilton in Bermuda and Gronnedal in Greenland.

The Eastern Atlantic Command (EASTLANT) based at Northwood in the UK comprised the Maritime Air Eastern Atlantic Area (MAIREASTLANT) based at Northwood; the Northern Sub-Area Command (NORLANT) based at Rosyth in the UK; the Maritime Air Northern Sub-Area Command (MAIRNORLANT) based at Pitreavie in the UK; the Central Sub-Area (CENTLANT) based at Plymouth in the UK; the Maritime Air Central Sub-Area Command (MAIRCENTLANT) based at Plymouth; the Submarine Forces Eastern Atlantic Area (SUBEASTLANT) based at Northwood; and the Island Commands of Iceland (ISCOMICELAND) and the Faeroes (ISCOM-

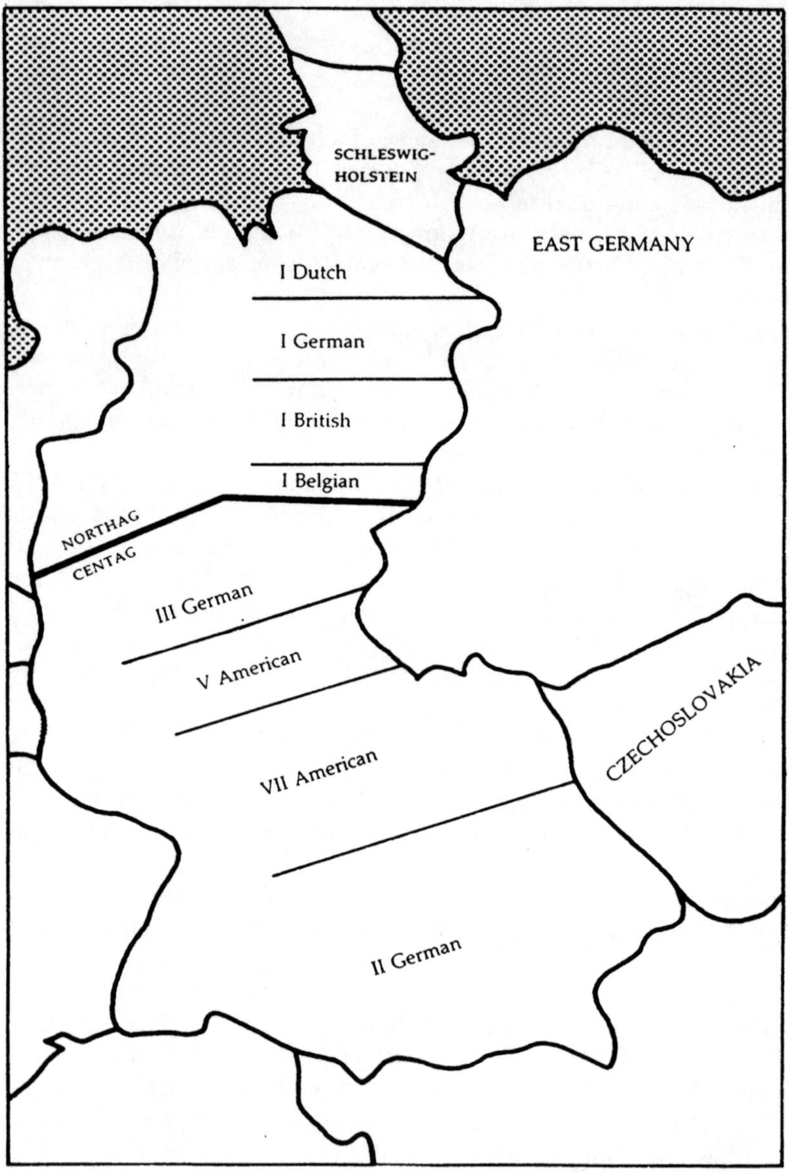

Figure 2.3 NATO Central Region Area Corps Sectors

Source: *NATO Facts and Figures* (Brussels: NATO Information Service, 1989).

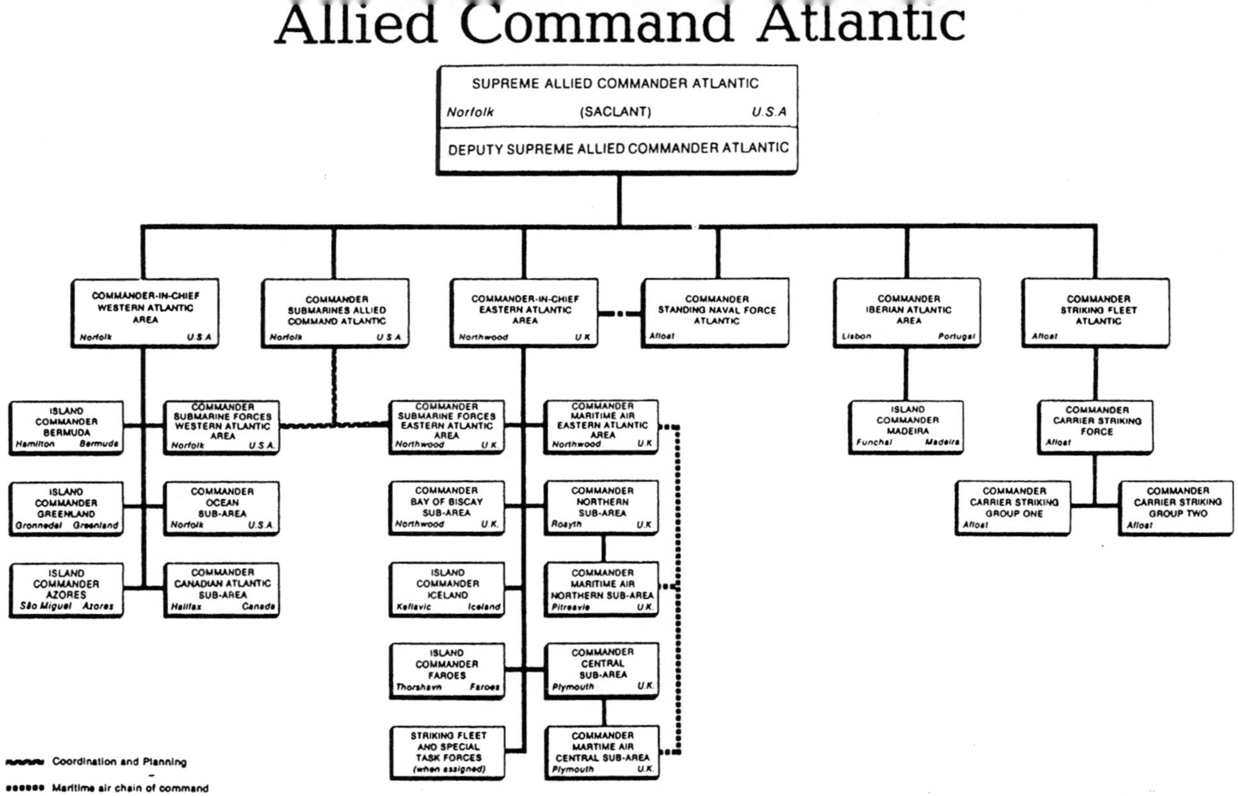

Figure 2.4 Allied Command Atlantic (ACLANT) Command Structure

Source: *NATO Facts and Figures* (Brussels: NATO Information Services, 1989).

FAEROES) respectively based at Kelfavik in Iceland and Thorshofn in the Faeroes.

The Striking Fleet Atlantic Command (STRIKFLTLANT) based afloat comprised the Carrier Striking Force (CARSTRIKFOR); the Amphibian Strike Force (AMPHIBSTRIKFOR), both based afloat; the Submarine Allied Command Atlantic (SUBACLANT) based at Norfolk Virginia; the Iberian Atlantic Command (IBERLANT) based at Lisbon in Portugal; The Standing Naval Force Atlantic (STANAVFORLANT) based afloat.[12]

The third major NATO command was the Allied Command Channel (ACCHAN) based at Northwood (see Figure 2.5). CINCHAN, the commander of ACCHAN, always a British officer, was responsible for the defence of the British Isles and the Channel region.

ACCHAN comprised seven Principal Subordinate Commands: the Benelux Sub-Area Channel Command (BENECHAN) based at Den Helder in the Netherlands; the Nore Sub-Area Channel Command (NORECHAN) and the Maritime Air Nore Sub-Area Channel Command (MAIRNORECHAN) both based at Pitreavie; the Plymouth Sub-Area Channel Command (PLYMCHAN) and the Maritime Air Plymouth Sub-Area Channel Command (MAIRPLYMCHAN) both based at Plymouth; the Allied Maritime Air Force Channel Command (MAIRCHAN) based at Northwood; the Standing Naval Forces Channel Command (STANAVFORCHAN) based afloat. ACCHAN operated British nuclear weapons and had a limited number of US nuclear weapons earmarked for use by the UK.

To facilitate the transfer from national to NATO command military personnel held both national and NATO commands simultaneously. SACEUR, for example, also held the post of CINCEUR (Commander-in-Chief Europe) a US national military command in which role he headed all US forces in the European theatre.

This duality of parallel national and NATO command structures was duplicated at every level in NATO and for almost all forces.[13] In peacetime national posts usually took precedence and forces were responsive to national control. The British officer who held the position of CINCHAN, for example, also held the national post of Commander-in-Chief Fleet and it was the latter which took precedence outside a NATO conflict.

In an effort to co-ordinate forces it was not unusual for officers to hold more than one NATO post responsible to more than one MNC, particularly in areas of command overlap (for example the waters

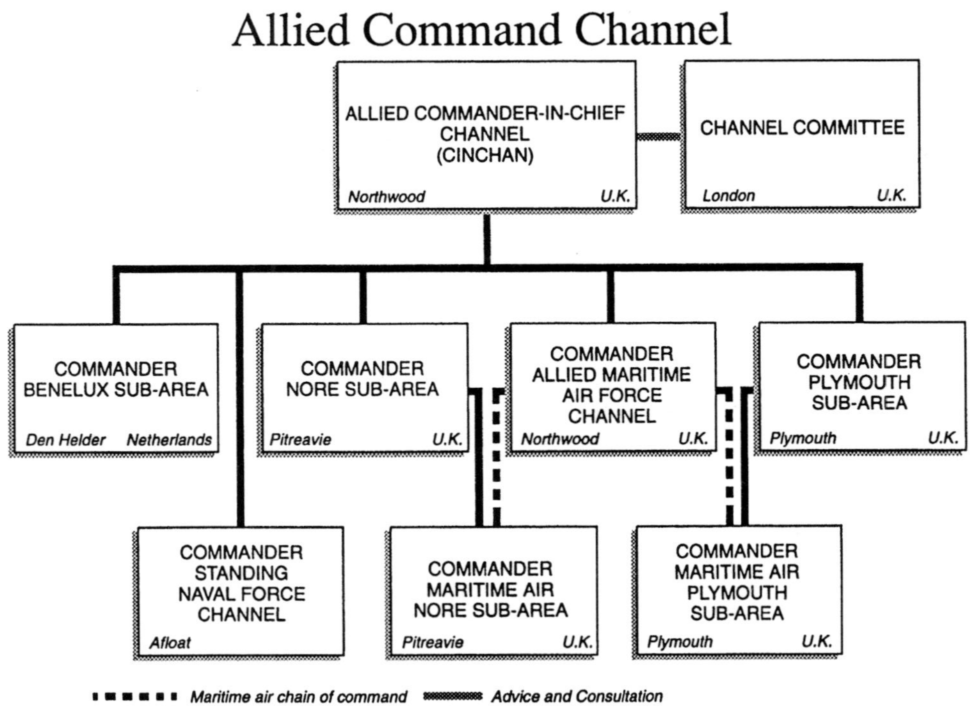

Figure 2.5 Allied Command Channel (ACCHAN) Command Structure

Source: *NATO Facts and Figures* (Brussels: NATO Information Services, 1989).

around the UK). Thus CINCHAN, to whom we have already refered, was also Commander-in-Chief of the Eastern Atlantic (CINCEASTLANT) a post subordinate to, and in the command chain of, SACLANT.[14]

NATO COMMAND AND CONTROL SYSTEMS

The task of linking the various elements within NATO required a command and control system of great complexity. This infrastructure had developed since the early days of NATO with newly introduced systems often overlaying, and providing redundancy and enhanced capabilities for, existing systems. The result of this evolutionary and largely *ad hoc* approach has been the development of a mass of systems and networks of varying age, sophistication, technical capability and survivability offering patchwork command and control in some areas and great security and redundancy in others.

In order to explore this tangle of systems and take from it an understanding of the strengths and weaknesses of NATO nuclear command and control it is necessary to simplify the discussion along thematic lines.

NATO's command and control infrastructure was broadly developed to meet command and control demands in three distinct but overlapping environments. The first was the politico-military environment between NATO member capitals, the political bodies of NATO (NAC, DPC, etc.), and the major NATO commands. The second was meeting the military command and control requirements within each of the three MNC areas of operations to assure command and control of all the forces at the disposal of each of the MNCs. The third was integrating the command and control capabilities of each of the three MNCs to try to enable ACE, ACLANT and ACCHAN to co-ordinate military operations at all levels of military activity.

The command and control requirements of the politico-military environment have been dealt with in some detail in the previous chapter's discussion of the NICS programme and its history. This section is concerned with the latter two environments: intra- and inter-MNC.

To consider intra-MNC command and control it is necessary to limit the discussion to only one of the three NATO MNC regions. Since

ACE has been the focus of NATO planning since its inception and since ACE operated the greatest concentration of nuclear command and control systems this will be the chosen theatre.

Within each MNC theatre command and control can further be broken down into four key elements. The first is communications, the second command, control and information systems (CCIS), the third intelligence and target acquisition, and the fourth air command and control systems (ACCS). The first three of these will form the strands of this section. The latter, ACCS, will be considered separately in the section below which discusses early warning and attack assessment.

Allied Command Europe controlled forces from all of the NATO member nations (except France)[15] placing an enormous burden on command and control. When SHAPE, the command headquarters of ACE, was first activated in 1951 the various countries which then comprised NATO were reliant on commercial PTT (Post, Telegraph and Telephone – shorthand for commercial communications systems) for much of their communications needs and on the remnants of Second World War tactical systems still deployed in Europe. As a consequence '. . . it often took hours to complete a call from some of the more distant HQ's under SHAPE's command'.[16]

The earliest attempt to provide dedicated military communications to meet the strategic needs of NATO was the development of the DOUBLE JUMP tropospheric scatter (troposcatter) system. A pilot DOUBLE JUMP scheme, known as project 'Hot-Line'[17] was operational in Norway by August 1958 and in the following year the first operational section was ready linking SHAPE (then in Paris) to subordinate HQs in Naples (Italy) and Izmir (Turkey).[18] DOUBLE JUMP itself was the initial phase of a more ambitious plan to link the European NATO members through the ACE-HIGH project.

ACE-HIGH was the first of the NATO, as distinct from national, funded command and control infrastructure programmes. It provided near real-time communications between SHAPE and all of the ACE PSCs covering a distance of over 9,000 miles.[19] The system relied on a mixture of troposcatter (over-the-horizon) and line-of-sight radio links and was fully operational by November 1963.[20]

ACE-HIGH remained one of the principal dedicated carriers of ACE theatre-wide communications. By 1990 it comprised 49 troposcatter and 40 microwave line-of-sight links supporting 570 voice, 260 telegraph and 60 data circuits.[21] Although these components had been modernised during the intervening years the system as a whole was largely, and in parts severely, outdated by 1990 and earmarked for replacement by systems based on linking national digital carriers across Europe.[22]

Aside from its age ACE-HIGH had another flaw in that it was a communications chain, the function of which was dependent on each link remaining operational. To go some way towards addressing this vulnerability the power of the troposcatter elements were boosted to allow jumping over faulty links. However, this had only limited utility because of the distances involved and the presence in the chain of microwave links which did not allow this facility.

The second programme linking ACE commands was based on satellites which offered greater survivability and connectivity than ACE-HIGH. Following the move of SHAPE to Belgium satellite terminals were installed at SHAPE and AFCENT which made use of the US Initial Defence Communications Satellite Program (IDSCP) satellites placed in orbit in 1966 and 1967.[23]

The NATO Satcoms discussed in Chapter One as part of the NICS project assumed this role when they were first deployed. In addition NICS offered a range of land-line, line-of sight, and other dedicated carriers which ACE used to link SHAPE to theatre MSCs and PSCs.

Links between SHAPE, the MSCs, and the PSCs down to the Corps level relied primarily on High Frequency (HF) radio in the theatre (since NICS does not operate below PSC level), the most important of which were the NATO HF Communications Network[24] and the CIP-67 analogue line-of-sight network.[25] NATO subsequently backed-up these important systems with a Shelterized HF Communications System delivered in late 1982 and operational by 1987. This provided 23 transportable HF systems – housed in nuclear weapons effects hardened shelters which could be deployed to remote sites across the ACE theatre – together with two fixed sites.[26] Across the theatre these systems were backed up still further by national PTT.

At or below Corps level NATO issued a directive which permitted the development of national rather than integrated NATO command and control systems.[27] A diversity of national systems consequently evolved both because individual nations favoured their national electronics companies for equipment and because NATO lacked an integrated approach to C^3 system architecture at this level. This approach created problems of interoperability which are explored in Chapter Six.

The second element of command and control within ACE was the investment in CCIS.[28] These were intended as information storage, processing and distribution systems. A CCIS usually had at least four roles: situation determination, including the status of friendly and enemy forces; situation evaluation, including data analysis in respect of objectives; planning, to consider alternatives for various operational scenarios; and implementation, the execution of plans, orders and instructions within the CCIS theatre.

The first CCIS introduced into NATO was the American World Wide Military Command and Control System (WWMCCS or WIMEX) used in the AFCENT region since 1975. (WIMEX will be discussed in the next chapter.) A number of NATO CCIS were subsequently introduced within ACE. The most important was the SHAPE CCIS which '. . . support[ed] the function of SACEUR and his staff' and was housed in the underground command centre beneath SHAPE.[29]

In the northern ACE region the Northern European Command CCIS (NEC CCIS) was deployed following initial planning in the late 1970s.[30] The NEC CCIS was itself composed of three sub-systems: the Allied Command Baltic Approaches CCIS (ACBA CCIS) which was established at Karup; the AFNORTH CCIS based at Kolsas which served the commander of AFNORTH and his staff; and the NOR CCIS which equipped the Norwegian headquarters and some air force bases in the region.[31]

In the southern ACE region (Greece, Italy and Turkey) there was only limited CCIS investment with the relevant commands fielding systems lacking the sophistication or capabilities of their northern counterparts. CCIS in the region were predominately introduced at commands below PSC level. The limited-capability CCIS at MEDCENT HQ in Santa Rosa (Italy), at MEDEAST located near Athens, and at MEDNOREAST located near Ankara were based on commercial equipment which allowed regional interoperability but at the price of reduced military hardness and reliability specifications. The first of the three networks was operational in 1985, the last in 1989.

The final CCIS within ACE was the UKAIR CCIS based at High Wycombe in the UK which was only partially operational by 1990.[32]

CCIS were also being introduced within ACE at and below the Corps level. As with communication systems at this command level NATO allowed national CCIS development with the consequent multiplication of different national systems. On the battlefield the US fielded ATCCS, the Italians CATRIN, the British Wavell, the West Germans HEROS, Canada TCCCS and the French SICF.[33] As with battlefield communications there were interoperability problems between the systems.

Within ACE there existed a long-standing shortfall in the area of intelligence and target acquisition. Until the late 1980s ACE had to rely on a patchwork of national systems which fed information to SHAPE through command channels. This resulted in system problems including information gaps, overlapping and conflicting information provision and slow information processing which hindered effective and timely decision-making by commanders.[34]

NATO made no attempt to standardise the target acquisition systems and intelligence systems within ACE. In the field of battlefield target acquisition, for example, the US completed trials with the JSTARS system, the British developed the ASTOR system, and the French fielded the Orchidee system, to name but three.[35] Whilst amongst intelligence systems the US fielded ASAS/ENSCE, the British Wavell/Gemini,[36] the Italians SORAO/CRESO, the French SATIS, and the West Germans Jasmin.[37] As a result NATO was forced down the path of attempting to co-ordinate national systems utilising computer processing technology to provide a common near real-time picture of the theatre for all users. This was the concept behind the Battlefield Information Collection and Exploitation System (BICES) project.

As the name implies BICES was a data fusion programme which involved Belgium, Canada, Denmark, France, Greece, Italy, the Netherlands, Norway, Spain, Turkey, the UK, the USA and West Germany. It was aimed at tying together intelligence and target acquisition information at or below the SHAPE command level.[38]

Although BICES was only at the feasibility study stage when flexible response was abandoned[39] important initiatives had already been taken in information integration at SHAPE. In 1986 NATO installed the War Headquarters Information Display and Dissemination System (WHIDDS) at SHAPE HQ.[40] WHIDDS provided detailed geographic information on the force deployments in the theatre by integrating information from a variety of intelligence inputs.[41] A WHIDDS system was also reportedly deployed at the NATO/UK HQ at Northwood on a trial basis in 1987.[42]

It was eventually intended that through BICES NATO would link together all theatre intelligence and target acquisition using the CCIS networks as information carriers and processors, but throughout the period of this study NATO had no such capability.[43]

Inter-MNC command and control, the third of the NATO environments introduced above, was intended to co-ordinate the commands and forces of the three MNCs.

In the area of communications the three MNCs were linked primarily by the NICS project systems already described in chapter one. Indeed the architecture of NICS itself arose from a plan, named the TRI-MNC, originally intended to link the MNCs more directly.[44] Similar levels of inter-MNC communications connectivity existed at the MSC and PSC levels. However, below this level inter-MNC communications were problematic because of interoperability limitations. Interoperability problems were also apparent between the MNCs in the areas of CCIS and intelligence and target acquisition.

NUCLEAR TARGETING

The targeting of nuclear forces in NATO required not only the co-ordination of the national targeting policies of the US and UK (and possibly France – see Chapter Five) but also the co-ordination of strategic and theatre forces to ensure the required coverage of targets and to minimise unnecessary duplication of targeting. The targeting process also provided another means by which NATO's European members could exert an influence on NATO's nuclear policy.

The co-ordination of targeting occurred at two levels: strategic and theatre. At the strategic level all US nuclear targets were co-ordinated at the Joint Strategic Target Planning Staff (JSTPS) based at Offut AFB in Nebraska, the headquarters of Strategic Air Command (SAC).[45]

When it was established in 1960 the JSTPS had two functions. The first was to provide the National Strategic Target List (NSTL) which described all of the strategic targets which the US needed to strike in the event of a nuclear war. The second was to provide the Single Integrated Operational Plan (SIOP) which assigned weapons to targets for all US strategic forces under a variety of employment scenarios.[46]

The last flexible response version of the SIOP, SIOP-6F, based on the National Strategic Target Data Base (NSTDB, which replaced the NSTL) included over 50000 target assignments (approximately five times as many as then available strategic weapons in the US arsenal). The SIOP itself was broken down into two main plans. The first was known as the Alert Response Plan (ARP) and set out a worse-case scenario in which only those strategic forces on routine alert were available and it assumed the US has suffered a nuclear attack. The second was known as the Generated Operations Plan (GOP) and set out a best-case scenario in which all US strategic forces were raised to alert level and used before an attack struck the United States.

The SIOP was further divided into four options: Major Attack Option (MAO); Selective Attack Option (SAO); Limited Nuclear Option (LNO); and, Regional Nuclear Options (RNO) which permitted targeting flexibility in a range of scenarios.[47]

In the UK British strategic forces were assigned to both national and NATO targets sets. National targets were determined by the Ministry of Defence (MOD) and NATO targets by the JSTPS (see Chapter Four for details). At the strategic level US and UK targeting were co-ordinated through the European Liaison Group to the JSTPS.[48]

At the theatre level nuclear targeting in NATO was determined by the Nuclear Planning Group (NPG), US European Command, NATO's International Military Staff (IMS), and the three MNCs. Operational plans based on the guidance of these groups were developed by the Nuclear Activities Branch at SHAPE.[49] Target plans were set out in the Nuclear Operations Plan (NOP) NATO's equivalent to the SIOP, formally known as SACEUR SUPPLAN 10001A (SACEUR Supplementary Plan).[50] The task of deconflicting the SIOP and NOP was the responsibility of the European Liaison Group and the JSTPS.[51]

The NOP provided for both selective and general nuclear response options. The general nuclear response option was further divided into a Priority Strike Program including time urgent targets such as enemy theatre nuclear forces and a Tactical Strike Program aimed at targets of tactical relevance but lacking the same time urgency such as logistic support elements.[52] In 1985 the NOP reportedly contained around 18500 targets of which approximately ten per cent were priority targets.[53]

The American and British SSBNs assigned to NATO were assigned targets from the SIOP rather than from the NOP suggesting perhaps that they were intended to be used only as part of a strategic nuclear option.[54]

At the battlefield level the location of enemy targets would be too fluid to permit detailed advanced planning of weapon to target allocation. Nevertheless a degree of political and high level military control of the targeting of these weapons was still maintained.

This was achieved by limiting in advance the parameters of the use of tactical nuclear weapons through the use of Selective Employment Packages (SEP). An SEP would not usually specify exact targets but would place limits on the type, numbers, timescale and other relevant variables of the use of the weapons.

In the process of involving the Europeans in nuclear planning in NATO, which was discussed in the previous chapter, the question of the size and nature of SEPs was one of the more contentious issues between the Europeans and Americans. The Europeans envisaged the initial use of nuclear weapons in discrete packages of single figure or perhaps a few tens of weapons as a maximum, while the US envisaged as many as 200 nuclear weapons being used in a single corps sector SEP.[55]

A US military document from 1976, although dated, provided some insight into the nature of an SEP planned by the US Army in Europe:

> The number of weapons in a Corps package will usually be from 100 to 200 depending on the threat, the mission, the terrain and population characteristics. A brigade might find as many as 20 to 30 weapons employed in its area or as few as 0 to 5, depending on the threat in its area relative to the rest of the corps area.[56]

Limitations of time also governed the use of SEPs and these had two dimensions. The first, the *timeframe*, was the period over which the Corps foresaw the need for nuclear weapons or command authorities authorised their use. Typically the timeframe extended from a period of hours to a day or more. The second, the *timespan*, refered to the period (or pulse) within which the SEP was to be executed. It was typically a far shorter time – of perhaps a few hours or less – intended to co-ordinate the use of weapons in the SEP and signal the controlled nature of their use.[57]

The flexibility of the SEP extended still further. The US Army specified five levels of the use of tactical nuclear weapons:

DEMO	Demonstration use of a very limited number of weapons with severe contraints on the location and nature of the use;
LDU	Limited Defensive Use
RBAU	Restricted Battlefield Area Use
EBAU	Extended Battlefield Area Use
TWU	Theatre Wide Use, the most general level of use with the least contraints across the theatre.

The political control of tactical nuclear weapons at sea was similar to that of battlefield nuclear weapons in that political leaders were unable to specify weapons and targets and instead relied on placing limits on the scale and parameters of use through procedures similar to those of SEPs.[58]

The final variable in NATO targeting was the targeting of French nuclear weapons which were not subject to formal NATO discussion. This is discussed in Chapter Five.

EARLY WARNING AND ATTACK ASSESSMENT

One of the more serious operational problems facing NATO was obtaining information and agreed evaluation of attack warning and assessment. For NATO to co-ordinate forces and activities required an assessment of threat which was agreed (however broadly) by all member nations. Yet NATO lacked the required supranational intelligence and early warning systems to enable this in all but a few mission areas.

For most early warning and attack assessment NATO relied upon the national systems of its member nations. In practice this meant a heavy reliance, to the point of dependence, upon the United States.

To discuss the elements of this dependence it is necessary to consider the two levels of warning and deal with each in turn. The first, strategic warning, can be defined as 'warning of impending attack prior to its execution'.[59] The second, tactical warning, can be defined as 'warning of an executed attack'.[60]

Strategic warning relied on intelligence systems which were able to monitor the activities of an adversary to determine whether preparations for attack were underway. NATO itself had no such systems and consequently relied on the USA which deployed a huge intelligence apparatus based on satellite and ground-based intelligence-gathering systems deployed around the world.[61]

As part of this apparatus the US had specific intelligence components for the European theatre which were part of the wider global apparatus.[62] Information from the US, and other NATO allies, was normally passed to the SITCEN at NATO headquarters and disseminated to the three MNCs.

Tactical warning addressed two threats. The first was from missile attack, the second from air-breathing vehicles (aircraft and cruise missiles). The early warning and attack assessment systems in place to address these threats for NATO were quite distinct and separate.

NATO was completely reliant on the US for early warning of missile attack. The US deployed a multi-element early warning apparatus as part of its own defence and information from this would have been passed to NATO.

The first element of early warning was the Defense Support Program (DSP) satellites in geostationary orbit around the world.[63] These satellites carried infra-red sensors which were able to detect the heat from the plumes of launching missiles. Information from the DSP satellites fed directly to the North American Air Defence (NORAD) complex in the United States and also to a few selected centres in the NATO theatre (see below).

The infra-red sensors on board the satellites could bring most of Europe and the Soviet Union under surveillance [with the exception of the United Kingdom with important subsequent implications], but were reported not to be able to scan more than one per cent of the total surveillance area at a given time. Although the DSP satellites were routinely trained onto Soviet missile fields with some accuracy there were area coverage limitations. Furthermore in the case of Soviet ballistic missile submarines, whose locations were largely unknown, the DSP had less utility.[64]

There were, moreover, technical limitations with the DSP satellites. According to one author the satellites '. . . basically just [gave] you an early indication that something is probably happening'[65] while another report suggested that they were able to '. . . estimate the approximate number of attacking missiles, their launch points, and their flight corridors, but could not determine the missile's destination'.[66]

In 1989 the US launched the first of a predicted nine improved DSP satellites intended to replace those then in orbit. The new satellites carried infra-red sensors with 6000 elements (three times the number on earlier versions) which were said to provide extended area coverage and improve target acquisition and assessment capabilities. The new satellites were thus reported to provide '. . . greater tactical missile warning capabilities in areas such as Europe'.[67]

The new DSP satellites promised limited improvements but did not offer NATO complete attack assessment and characterisation. They could, however, distinguish an attack aimed at the European theatre from an attack on the continental United States, and important advance on the previous system.[68]

The second system in US tactical warning of missile attack was the Ballistic Missile Early Warning System (BMEWS). BMEWS comprised three radar sites at Clear (Alaska), Thule (Greenland), and Fylingdales (UK), each with a range of over 3000 kilometres.[69] BMEWS was intended primarily as a US early warning capability and was configured accordingly. The radars consequently had only a limited utility for the European theatre. Information from BMEWS, as from the DSP, was passed to NORAD and from there was disseminated to SHAPE and other US and allied command centres. The UK also tapped information from the Fylingdales radar directly to British command centres (see Chapter Four). Because of this rather circuitous route for disseminating information to NATO question marks existed over the timeliness and value of BMEWS information for NATO.

Being ground-based BMEWS did not begin to detect missiles until they had risen some distance above the curvature of the earth, by which time incoming missiles could have been airborne for several minutes. Reportedly signals at the limits of BMEWS range needed to be reinforced over a period of time before missile detection could be confirmed, incurring further delay. According to one source it could therefore have taken as long as ten minutes for BMEWS to confirm attack and begin providing predicted impact location information.[70] Although this provided the US (at 30 minutes flying time distance from the Soviet missile fields) with up to 20 minutes warning there was no such margin for European targets. Furthermore, doubts about the utility of BMEWS for Europe were strengthened when one considered the inevitable additional delay in passing BMEWS information to NATO via NORAD (or possibly via the United Kingdom).

The BMEWS system underwent an upgrade programme in the late 1980s which began the process of replacing the existing radars with phased-array radars to eventually extend the target acquisition and attack assessment capability of the radars and, in the case of Fylingdales, extend the warning sector of the radar from 120 to 360 degrees. The improvements were intended to enhance technical capabilities but did little to address the doubts over the timeliness of BMEWS warning to NATO.[71]

The third US missile warning system of relevance to NATO was the Nuclear Detonation Detection System (NUDET). The updated version of the system known as the Integrated Operational NUDET Detection System (IONDS) was carried on 18 NAVSTAR/GPS navigation satellites in geostationary orbit. IONDS was reported to be able to detect the detonation of nuclear weapons with an accuracy down to 100 metres.[72] Intended as an aid to attack assessment by providing information about nuclear strikes and indirectly about surviving friendly forces the system could, in the early stages of a limited nuclear conflict in Europe, have provided information about the scale, nature, and locations of attack though again only by shunting information to NATO via NORAD.[73]

The second threat for which tactical warning systems were deployed was from air-breathing vehicles (aircraft and cruise missiles). NATO deployed its own multinational air defence system which extended the length of the theatre from north Norway to Turkey. Termed NADGE – NATO Air Defence Ground Environment – the system was developed during the early 1960s and first became operational in 1962.[74]

The modern NADGE which evolved over almost 30 years comprised by 1990 more than one hundred sites of radars, data processing and operation centres, and involved nine NATO nations.[75] The NADGE was developed in parallel with the British UKADGE (see Chapter Four) and French STRIDA (see Chapter Five) networks to provide theatre-wide air defence coverage. In the 1980s the network was interfaced with the Spanish COMBAT GRANDE system and there were improvements in the German ADGE (GEADGE) and Danish/Norwegian NADGE sectors.[76]

The purpose of NADGE was to link together early warning radars with fighter and ground-based air defence missiles across the NATO theatre to provide co-ordinated early warning and air defence. It was the shortfalls in NATO's ability to carry out this mission which prompted the developments in NADGE which took place from the late 1970s.

There were three main areas of air defence weakness which NATO sought to address. The first was the threat from low-flying aircraft which could fly beneath the detection sweeps of ground-based radars. NATO moved to address this threat by purchasing 18 Airborne Warning and Control Systems (AWACS) aircraft derived from the Boeing 707.[77] Implemented in 1982 the AWAC program deployed AWACS aircraft along NATO's central front from bases at Gielenkirschen in Western Germany (1982), Konya in Turkey (1983), Preveza in Greece (1985) and Tripani in Italy (1986). In addition there was an AWACS operating base at Oerland in Norway.[78] The AWACS command – NATO Airborne Early Warning Force (NAEWF) – was permanently controlled by NATO under the formal command of all three MNCs but under the operational control of ACE.[79]

The British initially went ahead with an indigenous AWACS system based on the Nimrod aircraft in the late 1970s before scrapping the project in 1986 and purchasing Boeing 707 AWACS which will not be ready until the mid-1990s (see Chapter Four).

The presence of AWACS in NADGE created the second problem which was how to integrate and deconflict the early warning information from the AWACS into a network originally designed only for information from ground-based radars and how to deconflict the two sources of information to ensure that all NADGE users were provided with a common picture of the air defence environment.

A programme termed AEGIS (Airborne Early Warning/Ground Environment Integration Segment) was established in the early 1980s to achieve this task. It involved standardising the downlink communications systems for the AWACS based on the US JTIDS (Joint Tactical Information Distribution System) communications system and providing enhanced processing capabilities to enable information to be integrated. The project was unfinished when NATO abandoned flexible response.[80]

A third problem involved the air defence of NATO but related more widely to the management of all aircraft in NATO air space. NATO had always lacked a properly integrated real-time means to monitor and control the movement of aircraft within the theatre. Such a concept would include not only NADGE but also IFF (Identification Friend or Foe) and the assessment of friendly forces. To address this shortfall NATO proposed in 1982 an Air Command Control System (ACCS) for the ACE theatre to be developed and deployed between 1985 and 2000.[81] In addition the ACCS concept was intended to interface ACCS with all other ACE Command and Control Information Systems (CCIS) and eventually with other CCIS outside ACE such as UKAIR.[82]

Grand in scale and vast in budget (estimated in 1989 at $40 Billion) the ACCS concept went through a feasibility stage before grinding to a halt. The ACCS development team was disbanded in July 1989 because of wrangling between contracting groups.[83] A year later a rethink of the project was still underway based on the same concept and meeting the same needs but requiring a 'change of system emphasis'.[84]

One final area of related interest concerns the threat of cruise missiles for which NATO originally did little planning. On the central front AWACS probably went some way towards addressing the threat, but elsewhere, notably in the US and UK, the threat of sea-launched cruise missiles renewed interest and investment in long-neglected Over-the-Horizon Backscatter (OTH-B) radar technology. The systems which underwent trials and deployment in the UK, the only OTH-B radars in NATO, are discussed in Chapter Four.

ALERT PROCEDURES AND TRANSITION TO WAR

NATO forces were responsive to three levels of operational command. The first, termed Full Command, refered to the command which national political leaders exercised over their national forces within NATO. At any time the member nations of NATO reserved the right to withdraw national forces from the alliance or direct those forces to specific purposes which may or may not accord with those of NATO or other NATO allies.

The second, termed Operational Command, refered to the control of military missions by NATO command authorities once control of operations had been passed formally to NATO.

In the transition from peacetime to wartime readiness the forces of NATO were earmarked for transfer from Full Command to direct NATO Operational Command in the NATO command structure described above. This process was known as the 'Chop' or as 'Chopping'.[85]

The third, termed Operational Control, refered to the autonomy of field commanders to carry out military operations once authorised by senior military and political leaders.[86]

The process of transition, and the subsequent readying of forces for combat, was largely governed by the NATO alerting process set out in the highly classified NATO Alert Plan.[87] This described a series of increasing military readiness conditions known as NATO Alert Conditions (LERTCONs) which set out the actions necessary to effect those conditions. There were five LERTCONs each of which corresponded to a specific state of military readiness:

LERTCON 5 – Peacetime Readiness
LERTCON 4 – Military Vigilance
LERTCON 3 – Simple Alert
LERTCON 2 – Reinforced Alert
LERTCON 1 – General Alert

There was some ambiguity about who held the authority to raise the alert levels within NATO. According to one source the alert level could be raised to LERTCON 4 by any of the three major NATO commanders (MNC) but not further without political authorisation from the NAC or DPC.[88] Another, however, suggested that the authority of the MNCs may have extended further:

> SACEUR . . . can move to the three higher rungs (LERTCON 3, 2 and 1) on his own authority, but is expected to consult first with NATO Defense Policy [sic] Committee.[89]

This omitted to say whether SACLANT and CINCHAN had similar authority or whether SACEUR could raise the alert level across NATO or only within ACE. In practice a degree of autonomy was probably available for each of the three MNCs up to LERTCON 4 for the forces within their command. SACEUR as the principal military commander could probably also have raised the level across NATO (a high degree of co-ordination was inevitable anyway). Above LERTCON 4 the three MNCs probably needed to seek political agreement for their actions across the theatre, unless the speed of events overtook the politico-military process in which case alert levels could be directly raised above LERTCON 4 by the three MNCs.

The five rungs of the LERTCON 'ladder' should not be understood as fixed operational packages which necessarily affected all forces. The LERTCON process was more flexible than this and allowed the raising of the alert of selected forces within NATO tailored to operational requirements. Moreover, some of the forces within each of the MNCs were at different alert levels during normal operational circumstances. Whilst in peacetime most forces stood at LERTCON 5, some were held routinely at higher levels. A good illustration of this was a number of US and NATO aircraft loaded with a total of approximately 150 nuclear weapons which were kept on quick reaction alert (QRA) ready for take-off at short notice at numerous bases across Western Europe.[90] QRA was itself a specialised form of alert within the LERTCON process.

Furthermore the NATO LERTCON levels had to be related to, or dovetailed with, national alert levels amongst the individual NATO member nations. The US, for example, employed a Defense Condition (DEFCON) alert procedure (which is examined in the next chapter) and sometimes raised and lowered the alert level of US forces within NATO in response to events outside the theatre, without necessarily altering the NATO alert level.

An example of this was provided during the Yom Kippur Arab-Israeli conflict of October 1973. The US placed its global forces on a DEFCON 3 alert level during the conflict as a political signal to the USSR.[91] Within NATO this included flushing US nuclear submarines from port and placing greater numbers of nuclear aircraft on higher alert (equivalent to QRA).[92] This occurred without prior consultation with the European allies causing considerable European alarm.[93]

Whilst the unilateral action of national forces, particularly US forces, was a potential engine for escalation in Europe it was more probable within the context of a European conflict that DEFCON, LERTCON and other national alert procedures would be synchronised and dovetailed to try to ensure co-ordinated action. However, since national forces would not be under NATO control at the outset of the process of raising the alert levels (since this was one of the main objectives of raising the alert levels) the possibility existed at the lower alert levels for a ragged and unsynchronised response amongst the differing national elements as some forces responded more quickly than others to the alert process.

Although the contents of NATO's Alert Plan remain classified an informed assessment of the likely operational elements of the alert levels was published.[94] This indicated that between LERTCON 5 and LERTCON 3 national forces in NATO would have been made ready for combat and attempts would have been made to integrate command and control. Only at LERTCON 2 would alliance forces be placed under NATO command (the Chop) and the NATO command structure have taken on its final wartime shape.

CONCLUSIONS

This chapter has outlined NATO's arrangements and systems for collective military operations. NATO's military command structure centralised command in just three Major NATO Commanders – SACEUR, SACLANT and CINCHAN – of whom SACEUR exercised the greatest military influence within the alliance, greater even than that of the Military Committee to which SACEUR was theoretically subordinate. That the three MNCs were always commanded respectively by two American and one British officer (as NATO's nuclear powers) reflected the dominance of the US in the European theatre and the regional autonomy of the United Kingdom in the CINCHAN area.

Even the most rudimentary description brings home the diversity, complexity and geographic distances involved in NATO's command structure. Understanding the nature of the command structure itself is sufficient to illustrate the considerable problems of distance, geography, language and history which were inherent in so diverse and populous an alliance as NATO.

Yet the description of the command structure also illustrates how NATO resolved many of these problems, tying together multinational forces and linking each to high-level centralised command, whilst at the same time preserving national integrity of forces at and below Corps level and enabling national air, land and (where relevant) sea forces to co-ordinate activities.

NATO's command structure was, nevertheless, hierarchical and pyramidal in composition building into NATO the vulnerabilities such a structure inherently carries. These weaknesses are discussed in later chapters to pinpoint many of the problems of the command and control of nuclear forces in a multinational alliance.

The discussion of NATO's command and control system focussed on three elements: communications, CCIS, and battlefield intelligence and target acquisition. The picture which emerged was broadly similar with respect to each. In each system area development was largely evolutionary and ad hoc resulting in a lack of overall systems architecture and system elements which vary greatly in their capabilities.

At and below Corps level the development of national systems (rather than NATO-wide systems such as ACE-HIGH) has led to the deployment of a multiplicity of communications, CCIS and battlefield intelligence and target acquisition systems which created problems in terms of interoperability and the military cohesiveness of the NATO alliance.

The assessment of nuclear targeting highlighted the close relationship between NATO and US strategic and theatre targeting and the dominance of the US at every level of NATO nuclear targeting. The discussion also made clear the limitations of political involvement in nuclear targeting particularly with respect to short-range nuclear weapons raising the notion that in NATO some nuclear weapons may have been under tighter political control than others.

The discussion of attack warning and assessment in NATO pointed to two nuclear threats to NATO forces: those of ballistic missiles and those of air-breathing systems (aircraft and cruise missiles). With respect to the former the European nations in NATO's integrated military command were evidently almost totally reliant on the US for early warning and attack assessment. The limitations of US systems for the European theatre and the timeliness of information from those systems for NATO raised important questions about the utility of these systems for NATO warning.

With respect to air-breathing threats NATO's air defence network was genuinely theatre-wide and multinational. The systems involved were nevertheless not without problems – notably in the integration and management of information within and between systems – but NATO's air defence could generally have been considered sound.

The final part of this chapter introduced NATO's alert procedures. It described the extent of military autonomy with respect to raising the alert status and the way in which plans for the transition to war in NATO were shaped (but not automatically governed) by the preplanned activities envisaged by the specific steps in the alert ladder between LERTCON 5 (Peacetime Readiness) and LERTCON 1 (General Alert).

3 US Nuclear Command and Control in NATO

INTRODUCTION

With the exception of the relatively few British nuclear weapons committed to the alliance the United States was the nuclear armourer of NATO. It supplied not only the bulk of nuclear weapons for the alliance but also all of the nuclear weapons intended for use by the NATO allies through bilateral nuclear-sharing agreements. Understanding the command and control of these weapons is consequently pivotal to understanding nuclear command and control in the NATO alliance as a whole.

This chapter attempts to provide this understanding. It is organised into five sections. The first looks at the political control of US nuclear forces focusing on the dominant role of the President. The second considers the strategic command and control apparatus which was in place to support his authority. The third discusses the connectivity between the National Command Authority (NCA) and US nuclear forces in Europe and looks at the only US strategic force permanently assigned to NATO, namely the ballistic missile submarines. The fourth looks in detail at the arrangements for the use of US theatre nuclear weapons in Europe by US forces. The fifth considers the command and control of US theatre nuclear forces in Europe which were at the disposal of bilateral NATO forces.

POLITICAL AUTHORITY

Under American constitutional law the command of US nuclear forces was explicitly placed in the hands of the US President:

> Under existing law, the President alone has the basic authority to order the use of nuclear weapons. This authority, inherent in his constitutional role as Commander-in-Chief, may be delegated to subordinate officers in the chain of command virtually without

> limitation . . . The President's authority to order the use of theater nuclear weapons in the event of a war involving NATO, while subject to certain procedural arrangements, is similarly unlimited.[1]

Although Presidential control over US nuclear weapons was formally absolute, in practice it was modified, to some degree, by the limitations of the President's role in US government and his relationship to the Legislative and Judicial branches of government,[2] by the limitations of his role as Commander-in-Chief,[3] and by his commitment to consultation with allies, each themselves subject to certain caveats.

Although the President had the sole authority and responsibility to order the use of nuclear weapons the control of nuclear weapons operations was exercised by the NCA which, according to the 1971 Department of Defense Directive 5100.30, consisted only of 'the President and the Secretary of Defense or their duly deputised alternates or successors'.[4] The NCA could also be widened to include the most senior US military officer the Chairman of the Joint Chiefs of Staff (CJCS) and possibly the other Joint Chiefs.[5] The sharing of operational control – if not authority – enabled even this most senior level of decision-making to meet the 'two-person' rule which governs all US nuclear weapons activities. This rule stated that all decisions, procedures or processes involving nuclear weapons had to be carried out by at least two individuals, and was intended as a hedge against unauthorised or irrational action by anyone in the command chain.

The President exercised his authority over nuclear weapons through the possession of electronic codes which provided the authorisation to use nuclear weapons and the means to enable (i.e. allow to function) the weapons at his disposal. In the event that he needed to authorise the use of nuclear weapons the President would pass the codes – termed the Emergency Action Message (EAM) or 'go codes' – to subordinate commands for onward transmission to the nuclear forces.[6]

The EAM could be passed to selected forces for limited nuclear options or to all forces for general response. To provide the President with flexibility he was accompanied at all times, irrespective of his location, by a military aide (usually a warrant officer or equivalent) carrying a briefcase known as the 'football'.[7] The football contained a 75 page black book detailing the SIOP nuclear options available to the President, the EAM codes, a list of the classified sites to which the President could be taken for safety in an emergency and a description of the procedures for the Emergency Broadcast System (EBS) by which the President would address the American nation.[8]

The President was also linked at all times to systems which allowed him to communicate rapidly with the White House and indirectly with command centres and nuclear forces, again irrespective of his location. One such system was the global Mystic Star high frequency (HF) radio system which used the SAC GIANT TALK/Scope Signal III radio links as a carrier.[9] The President was thus, at least in theory, able to order the use of US nuclear weapons from anywhere in the world and at any time.[10] There were neverthless grounds – not least the complexity of the football's contents, the lack of Presidential familiarity with its contents, and the isolation from the NCA and senior military aides – for arguing that in some circumstances the President could find it difficult to exercise these options.[11]

In the event of the death or incapacitation of the President there existed a succession list, established by the 25th Amendment to the US Constitution, which set out, in order, those deputised to assume his office and authority. The list was as follows:

1 Vice President
2 Speaker of the House of Representatives
3 President pro tempore of the Senate
4 Secretary of State
5 Secretary of the Treasury
6 Secretary of Defense
7 Attorney General
8 Secretary of the Interior
9 Secretary of Agriculture
10 Secretary of Commerce
11 Secretary of Labour
12 Secretary of Health and Human Services
13 Secretary of Housing and Urban Development
14 Secretary of Transport
15 Secretary of Energy
16 Secretary of Education

To provide for the succession of the National Command Authority a second list established by President Eisenhower under Presidential Directive 10820 of 18 May 1959 provided for the succession, again in order, of the office of Secretary of Defense:

1 Deputy Secretary of Defence
2 Secretary of the Army
3 Secretary of the Navy

4 Secretary of the Air Force
5 Undersecretary of Defence for Research and Engineering
6 Assistant Secretaries of Defence and the General Counsel to the Department of Defence in order of length of service
7 Under Secretaries of the Army, Navy and Air Force in order of length of service
8 Assistant Secretaries of Army, Navy and Air Force in order of their length of service

Finally, the military element of the NCA reportedly also had a succession list to provide for continuity of the CJCS:

1 Commander-in-Chief Strategic Air Command (CINCSAC)
2 CINC NORAD
3 CINC Atlantic (CINCLANT)
4 CINC Central Command (CINCCENT)
5 CINC Europe (CINCEUR)
6 CINC Military Airlift Command (CINCMAC)
7 CINC Pacific (CINCPAC)
8 CINC Readiness Command (CINCREAD)
9 CINC Southern Command (CINCSOUTH)[12]

The Presidential chain of succession was, however, unlikely to allow a smooth transfer of authority if the President and NCA were incapacitated by some means because other members of the command chain were also likely to be lost, and because few of those on the list had an awareness of their place on it[13], and fewer still were sufficiently well briefed to do more than follow the advice of surviving elements of military command.[14]

In practice it would seem that in the event of the incapacitation of the President in the context of a crisis or potential nuclear conflict authority to use nuclear weapons would have passed to the Secretary of Defense, and on to the CJCS and subordinate nuclear commanders such as CINCSAC.

Thus far the discussion has refered to delegation of nuclear authority by force of circumstance but it was equally possible for the President to pre-delegate release and use authority at some point in a crisis, probably before the US suffered attack, to circumvent the potential for decapitation (that is the loss of the political leadership).

There is evidence that such pre-delegation took place even in peacetime. A US Congressional report on the control of nuclear weapons published in 1976 noted:

> . . . no . . . US or NATO field commander has been delegated authority to use [US] nuclear weapons without express approval of the President of the United States, with one exception . . . the North American Air Defense [NORAD] commander, who has been delegated such authority only under severe restrictions and specific conditions of attack.[15]

Another author claims that US President Eisenhower pre-delegated authority to use nuclear weapons – under certain circumstances – to his Pacific Commander (CINCPAC) in 1958 and argued that subsequent Presidents may have also delegated authority to overseas CINCs.[16]

Furthermore, in 1960 the commander of Strategic Air Command is also reported to have indicated that he possessed delegated authority to use nuclear weapons under certain circumstances.[17]

What is unclear is how long such standing pre-delegation persisted and whether, for present purposes, pre-delegation was ever extended to the US Commander-in-Chief, Europe (CINCEUR) the officer who also holds the post of SACEUR. NORAD lost in the late 1970s the formal, if limited, pre-delegation authority. In the absence of contrary evidence it seems improbable that other CINCs had routinely pre-delegated authority, however limited.

This is largely of academic interest since in a period of international tension which would almost certainly precede any conflict there would be ample time for the President to delegate authority.

A second area of potential circumspection of the Presidential authority over the use of nuclear weapons lies in relations with allies. Membership of NATO limited the President's authority (on a voluntary basis) over US nuclear weapons deployed in Europe in ways which were described in Chapter One. In practice the President may have been willing to seek the advice of allies prior to using nuclear weapons, but may have chosen to ignore the advice, may have found that 'time and circumstance' did not permit consultation, or may simply have chosen not to consult. No allies exercised a veto over US nuclear operations in NATO, even of weapons stored on allied territory.[18]

The ally with whom the US had the closest military ties and to whom the strongest commitment to consultation were made was the United Kingdom. Yet even here there was nothing mandatory or binding about the process. Aside from the NATO consultative process to which both were party the US and UK made a series of bilateral agreements governing the use of their respective nuclear forces and commiting each, nominally, to consultation with the other prior to their use in any theatre. For the UK these agreements went further to cover the deployment of US military forces (including nuclear weapons) in the UK.

The close co-operation between the two countries dated back to the 1943 Quebec Agreement (formally known as the 'Articles of Agreement governing collaboration between the authorities of the USA and UK in the matter of tube alloys' – the latter a coded reference to nuclear weapons) which committed each party to the undertaking that each '[would] not use atomic weapons against other parties without prior consultation with each other'.[19]

After the war this commitment was toned down to eliminate the nominal veto the commitment provided to either party over the actions of the other. In January 1948 a '*Modus Vivendi*' was reached which provided increased US nuclear co-operation with the UK's nuclear weapons programme but at the price of surrendering the UK's nominal veto even over US nuclear missions launched from the UK.

In April 1950 a formal arrangement, known as the Ambassadors' Agreement, which agreed the permanent basing in the UK of the US air force (including nuclear weapons) failed to consider the question of consultation or consent.[20]

It was not until January 1952 and the Truman-Churchill communiqué, following lengthy discussions, that the situation concerning US and UK consultation was rationalised. The communiqué stated:

> Under arrangements made for the common defence the US has the use of certain bases in the UK. We reaffirm the understanding that the use of these bases would be a matter for joint decision by His Majesty's Government and the US Government in the light of circumstances prevailing at the time.[21]

This communiqué remained the basis upon which the question of US consultation was resolved throughout the period of this study. Undoubtedly the phrase 'in the light of circumstances prevailing at the time' could be interpreted as a 'get out' clause irrespective of assurances from both parties to the contrary.

This ambiguity resurfaced twice in the 1980s once over the issue of the deployment of US nuclear cruise missiles in 1983 and again following the US attack on Libya in 1986 in which US bombers flew from British bases.[22]

The deployment of cruise, which as a NATO nuclear system is most germane to this discussion, re-opened the debate about the control of US nuclear forces based in the UK and consultation on their use.

At the height of this debate in October 1983 the then British Secretary of State for Defence Michael Heseltine refered to the Truman-Churchill communiqué and insisted that he had 'absolute assurances' that no US-made nuclear missiles would be 'fired or launched' from British territory without the permission of the British Prime Minister.[23]

If Heseltine was correct then at least for those nuclear weapons based in the UK the power of the US President was circumscribed. If the caveat in the 1952 communiqué was nearer the truth then there was no guarantee that US nuclear weapons could not be used from the UK without British permission and rather less limitation on the President's authority.

STRATEGIC NUCLEAR COMMAND AND CONTROL

Assuring the transmission of the EAM and SIOP options, which was essential for assured retaliation, necessitated both the presence of the President and NCA and secure communications to link the President and NCA with the nuclear forces. The first element in the command and control apparatus to support US nuclear forces were the command centres which were intended to guarantee the safety of the President, NCA and senior military advisers.

The White House, from which the President would normally conduct crisis management, was not hardened against nuclear attack. There was a bomb shelter beneath the east wing but this could not survive a nuclear strike, provided only limited communications facilities and could not function well as a crisis centre.[24]

The Secretary of Defense and the Joint Chiefs of Staff based in the Pentagon had at their disposal a primary command centre, termed the National Military Command Centre (NMCC), also within the Pentagon. The NMCC provided a full range of command and control capabilities and was linked to nuclear forces and senior US military commanders around the world. However it occupied office space on the second and third floors of the Pentagon and was therefore vulnerable to attack.[25]

Supporting and duplicating the functions of the NMCC was the Alternate National Military Command Centre (ANMCC) located beneath Raven Rock mountain in Pennsylvania.[26] The ANMCC was hardened against nuclear attack and could probably have survived all but a close or direct hit. It was possible, and in certain scenarios intended, that the President would transfer to the ANMCC in a crisis both for safety and for ease of decision-making if joined by the Secretary of Defense and members of the JCS.

Presidential safety was ultimately provided by an airborne command post, known as the National Emergency Airborne Command Post (NEACP pronounced 'kneecap'), based at Grissom AFB in Indiana.[27] Derived from a Boeing 747 the NEACP (designated E-4B in its militarily hardened form) hosted a command centre and huge variety of communications linking the President once aboard directly or indirectly to all US nuclear forces.[28]

It was envisaged in an emergency that the President would be taken by helicopter from the White House lawn (assuming he was in Washington) to Grissom AFB to board the NEACP. Once airborne the aircraft could sustain continuous flight for up to 72 hours with inflight refueling or could land for refueling and maintenance at secret bases.[29] It was intended that the President could direct US nuclear retaliation from the NEACP even if nuclear weapons were striking targets in the USA. To achieve this the NEACP had both direct communication links to some nuclear forces and links to other airborne command elements which in turn were linked to the remaining nuclear forces.

The most important of these support aircraft were the Post Attack Command and Control System (PACCS) also known as 'Looking Glass' or 'Cover All'.[30] The PACCS, based on the EC-135, a derivation of the Boeing 707, was a programme of continuous airborne alert which provided for at least one aircraft to be permanently airborne at all times by rotation (one aircraft taking off and assuming operations before the one it replaces lands) to provide invulnerable connectivity to nuclear forces. PACCS aircraft acted as an alternative headquarters for Strategic Air Command (SAC), which was based at Offutt AFB, and were able to control US strategic nuclear forces.[31]

The second important airborne command posts were the TACAMO (an acronym for 'Take Charge and Move Out') aircraft of the US Navy which provided connectivity with US ballistic missile submarines (SSBN).[32] Originally based on the EC-130 Hercules the TACAMO programme underwent an upgrade to E-6A designated aircraft based on the Boeing 707 in the late 1980s.[33]

NEACP, PACCS and TACAMO aircraft were linked by a variety of communications systems from UHF to VLF via the FLTSATCOM satellites, line-of sight radio links and relay aircraft, and direct communications to ground-based strategic nuclear forces (the missile silos), air-borne forces (SAC bombers on airborne alert) and sea-based forces (SSBNs).

Communications between aircraft downlinks and ground-based strategic nuclear forces were strengthened by the Ground Wave Emergency Net (GWEN) a nodal communications network intended to provide high survivability against even nuclear attack.[34]

Finally the US deployed the Emergency Rocket Communications System (ERCS) which consisted of ten Minuteman missiles based at Whiteman AFB whose front-ends contained transmitters rather than nuclear warheads. Launch orders for the ERCS could be issued from the ground or directly from the NEACP or PACCS aircraft (if authorised). As the missiles rose after launch they were intended to transmit blanket release codes and authorisation to all US strategic forces remaining intact at that point.[35]

Collectively the NMCC, ANMCC, NEACP and other principal command centres (including SAC headquarters) were known as the National Military Command System (NMCS). The centres, systems and infrastructure supporting the NMCS – which include the PACCS and TACAMO aircraft, the ERCS, satellite communications systems (FLTSATCOM and AFSATCOM), and ground-based systems (GWEN, the Survivable Low Frequency Communications System SLFCS, and other communications systems from UHF to VLF) – together constituted the Minimum Essential Emergency Communications Network (MEECN). MEECN included only those systems which provided hardened or high security command and control for the strategic forces.[36]

The NMCS and MEECN assets were themselves elements of the far larger World Wide Military Command and Control System (WWMCCS pronounced 'Wimicks' or WIMEX) established in October 1960 to 'provide the constituted authorities with the information needed for accurate and timely decisions and the reliable communications needed to transmit those decisions to the military forces under all conditions of peace and war'.[37]

COMMAND AND CONTROL OF US STRATEGIC FORCES IN EUROPE

US forces in Europe were under the command of the Commander-in-Chief Europe (CINCEUR) the officer who also held the post of SACEUR. Under national auspices CINCEUR commanded all US forces within the European theatre and in theory could order any military action within the theatre using purely national forces. Within NATO however these forces were assigned to SACEUR and subject to NATO multinational decision-making. The US capacity for unilateral action was in practical terms, however, undiminished.

Command and control of US nuclear forces in Europe was exercised by a variety of systems which addressed two main functions. The first was to provide assured trans-atlantic connectivity between the USA NCA and US nuclear forces in Europe. The second was to provide command and control of US forces within the theatre. This section is concerned with the systems which tied the NCA to CINCEUR and with the specific elements of US nuclear forces in Europe assigned a strategic role, namely the US ballistic missile submarines (SSBNs).

The first dedicated transatlantic military communications system was the North Atlantic Radio Relay System (NARRS) which extended from Canada to the UK and became operational in the early 1960s.[38] This system and the commercial carriers then available provided US forces with only limited and often unreliable communications and in the early 1960s work began on two military systems which were to greatly improve the situation (NARRS continued to link North America to Europe as a carrier for BMEWS information from Fylingdales in the UK to NORAD). The first of these was the AUTOVON (Automatic Voice Network) which provided a direct dial voice communications system. AUTOVON switching centres were built in Europe at ten sites: Hillingdon and Martlesham Heath in the UK; Feldberg, Lanerkopf, Donnesberg and Schoenfeld in Western Germany; Humosa in Spain, Coltano and Naples in Italy; and Athens in Greece.[39] AUTOVON was fully operational in Europe by 1969 and provided a number of facilities including conference calls, a direct 'hot-line', priority override for a limited number of subscribers, and automatic switching in the event of damaged or degraded circuitry.

The second of the systems was the AUTODIN (Automatic Digital Network) which provided message and data, rather than voice, transmissions. AUTODIN switching centres were constructed at three principal switching centres in Europe: Coltano in Italy; Pirmasens in Western Germany and Croughton in the UK.[40] There were also subsidiary switching centres at the headquarters of: NORTHAG/2ATAF; AFNORTH/COMSONOR; CENTAG; 4ATAF; AFCENT/AAFCE; LANDSOUTH/5ATAF; LANDSOUTH/6ATAF; AFSOUTH/AIRSOUTH/NAVSOUTH; AFSOUTH; and SHAPE.[41] AUTODIN also became fully operational in Europe in 1969.[42]

More recently elements of the AUTOVON network were upgraded by the AUTOSECVON (Automatic Secure Voice Network) which first became operational in the late 1970s. As the name suggests AUTOSECVON provided greater security voice communications although the system relied mainly on AUTOVON swicthing centres as long-haul carriers.[43]

Both AUTOVON/AUTOSECVON and AUTODIN were gradually replaced in upgrade programs which were incomplete by 1990. AUTOVON was being replaced by the Defense Switched Network (DSN). The DSN was in fact based on computer software packages which could be used on commercial PTT networks to provide the necessary military security and connectivity. By paying commercial telecommunications companies for the use of their circuits in this way the US Department of Defense intended to avoid the cost of purchasing new or upgrading existing dedicated networks. Such an approach was only possible because of the greatly improved systems provided by commercial operators.[44]

In 1982 the US DOD took the decision not to proceed with the planned AUTODIN II upgrade program and instead to develop the more advanced Defense Digital Network (DDN). Based on the ARPANET[45] the DDN was intended to provide greater traffic capability and enhance survivability. Eventually the DDN was intended to comprise 200 network nodes globally serving more than 750 computers and 30000 terminal users.[46]

Within the European theatre AUTOVON/DSN functions were supplemented by the huge US European Telephone System (ETS). First proposed in 1978 the ETS was a theatre-wide digital telephone system with 150 switching centres which eventually served 130000 military users in Western Germany (including Berlin), the Netherlands, Belgium, and Italy.[47] The system was operational by the mid-1980s but additional switches and user stations were added later before the system reached full specification. Although intended primarily for day-to-day peacetime operations the system did have wartime application and featured conferencing and priority override for a limited subset of users.[48]

Between 1979 and 1983 the US also deployed the Digital European Backbone (DEB) which was reportedly '. . . a microwave relay system which [was] . . . the primary defense communications system for the US Armed Forces in Europe'.[49] DEB had links throughout Belgium, Western Germany (and Berlin), Italy, the Netherlands, Spain, Turkey, and the United Kingdom. It was, as the name implies, a hardened system based on microwave relays (not unlike GWEN) which was intended as a high-security carrier for command and control information possibly including EAM transmissions.[50]

Further links were provided by the Defense Satellite Communications System (DSCS pronounced 'discus'). In 1967 the DSCS became the first satellite communicatons system introduced into Europe (see Chapter Two). It linked Washington to embassies around the world and to major military commands such as USEUCOM providing a high rate data and voice carrier.[51] In Europe there were many DSCS ground terminals, the principal military ones including: Kester in Belgium; Thurso, Coughton, Menwith Hill and Oakhanger in the UK; and Pirmasens and Landstuhl in Western Germany.[52]

The upgraded DSCS II consisted of four satellites in geosynchronous orbit over the Atlantic Ocean, Indian Ocean and the east and west of the Pacific, together with two reserves waiting in orbit in case of system difficulties. The DSCS was also important as the carrier of AFSATCOM relay terminals. AFSATCOM provided rapid high security data transmission linking the NCA to US strategic nuclear forces (AFSATCOM relay terminals are also carried on FLTSATCOM and SDS satellites).[53]

The AFSATCOM terminals were also one of the systems which linked the NCA to the SILK PURSE airborne command post aircraft based at Mildenhall in the UK since 1965.[54] SILK PURSE was the USEUCOM equivalent of the PACCS and was intended as a survivable headquarters for the CINCEUR capable of directing US nuclear forces in Europe from the air even during a nuclear conflict. Based on four EC-135H aircraft (derived from the Boeing 707) the SILK PURSE program was able to maintain at least one aircraft on permanent airborne alert during a crisis (though this was not attempted during in peacetime status). SILK PURSE carried a host of communications systems which, in addition to AFSATCOM links connecting the aircraft to NEACP, PACCS and TACAMO, included downlinks to European Ground Entry Points (GEP) enabling the aircraft to access US theatre communications systems, and a VLF transmitter for connectivity with US submarines.[55]

The most important of the networks was, however, the World Wide Military Command and Control System (WWMCCS). WWMCCS can be understood as a collective system which incorporates many of the systems thus far discussed. According to one source: 'Established in

October 1962 WWMCCS aimed to tie together and co-ordinate the quasi-independent theater and strategic systems of all the major US military commands in . . . a "system of systems"'.[56]

As defined by the JCS WWMCCS consisted of five functional elements: early warning, communications, command facilities, data collection and processing and executive aids.[57] The US early warning and attack assessment elements were discussed in the previous chapter and the communications and command centre facilities included both the NMCS and MEECN discussed above (the latter can be considered the essential backbone of WWMCCS). For the purposes of this discussion of transatlantic connectivity the remaining functions of WWMCCS can be considered as a global Command and Control Information System (CCIS).

WWMCCS was first introduced into Europe in 1975 at four sites: Northwood near London (UK) the headquarters of the US Navy Europe (USNAVEUR); Heidelberg in Western Germany the headquarters of the US Army Europe (USAREUR); Ramstein in Western Germany the headquarters of US Air Force Europe (USAFE); and Stuttgart-Vaihingen in Western Germany the headquarters of US European Command (USEUCOM).[58]

Until the early 1980s these sites functioned as regional CCIS for each of the three branches of US forces in Europe and USEUCOM. Information could be passed between the centres but not directly between computer systems. In 1983 the US undertook an upgrade termed the WWMCCS Intercomputer Network (WIN) based on the ARPANET to link the computers at all WWMCCS sites to allow global information exchange directly between computers.[59]

During the later 1980s WWMCCS was further upgraded in the WWMCCS Information System (WIS) program designed to more fully integrate computer and communications between WWMCCS sites and to use the Local Area Network (LAN)[60] concept to enhance system survivability.[61]

Before looking specifically at the command and control of US nuclear forces in Europe there is one more network which needs to be discussed: the GIANT TALK/Scope Signal III system. GIANT TALK was a global system of HF stations which linked the NAC to SAC nuclear bombers via ground stations. Scope Signal III was the HF replacement project for GIANT TALK designed to provide enhanced survivability and greater security. There were at least four GIANT TALK stations in Europe: at Torrejon in Spain; Incirlik in Turkey; and

Barford St John/Croughton and Mildenhall in the UK,[62] part of a reported network of 16 stations worldwide.[63]

There is considerable ambiguity about whether all GIANT TALK sites were upgraded to Scope Signal III, but those in Europe, which are of primary concern to us here, were upgraded.[64]

US strategic nuclear forces formally assigned to NATO comprised the Poseidon SSBNs based at Holy Loch in the UK. Although nominally under the command of SACEUR, operationally they were controlled by SACLANT from Norfolk (Virginia, USA).[65] Command and control of Poseidon was exercised through SACLANT and used the VLF transmitter at Cutler in Maine (USA) as the principal means of communications.[66] This was supported by other VLF transmitters in the USA and by the TACAMO fleet. It was also possible that orders could have been passed to the submarines via the NATO VLF transmitters in Europe at Travolara in Sardinia and Anthorn in the UK.[67] SILK PURSE's VLF transmitter could have provided additional connectivity in an emergency. The US also experimented with ELF transmitters during the 1980s and in 1984 demonstrated the ability to communicate with a submarine in the Mediterranean using the ELF transmitter in Michigan state.[68]

Finally, it is possible that the national-owned VLF transmitters at Criggion and Rugby in the UK (see Chapter Four) and at Helgeland in Norway could also have been used to communicate with American submarines. Although if US national systems were not available these were hardly likely to be available either.

COMMAND AND CONTROL OF US THEATRE NUCLEAR FORCES IN NATO

US theatre nuclear forces in NATO were deployed mainly in secure storage facilities across the theatre. The exceptions to this were those weapons routinely deployed on naval vessels and those relatively few weapons kept until 1990 on QRA alert. In all three situations extensive efforts were made to ensure the security of weapons, monitor their numbers and movements, and, if required, prepare, facilitate and exercise their use.

Control was imposed primarily by the use of Permissive Action Links (PALs). PALs can be understood as electronic locks built into nuclear weapons systems (warheads or launch control facilities) which require the correct electronic coded 'key' to unlock them and enable their use. Without this key a weapon will not function even if it can be launched or fired. PALs were first developed in the early 1960s specifically in response to the need for control of American nuclear weapons deployed overseas and were first installed on nuclear weapons in 1962.[69]

The first Category A PALs have been described as:

> . . . simple electronic locks. When the correct numerical code was fed into the PAL device the electrical circuit of the nuclear weapon was completed and the weapon could be detonated.[70]

Category A PALs were first installed on QRA aircraft in Europe and on a range of weapons then under dual US-NATO ally control under bilateral agreements (these included the Jupiter, Mace, Pershing I, and Sergeant missiles, though not the Thor which was withdrawn from Europe at the same time as PAls were first introduced).

Over the next twenty years PAL technology improved markedly become increasingly sophisticated and far less easy to circumvent or tamper with. In all there were five versions of PALs to 1990: Category A, B, C, D, and F.[71] The later Categories D and F (there has been no 'E') offered a range of functions:

- 6 or 12 digit (respectively) electronic coded lock.
- multiple code feature: this is '. . . similar in concept to locks that have individual, submaster and master keys allowing single weapons, or classes or weapons to be unlocked with different codes'.[72]
- limited-try facility: this means that the operator had only a limited number of attempts to insert the correct code before the weapon locks itself irretrievably and could not be used. Unauthorised attempts to unlock the PAL by the multiple entry of random numbers quickly rendered the weapon inoperable.

The importance of PALs within the European theatre is hard to overstate. The devices provided a high degree of security and assured high level control of nuclear weapons by the NCA. Presidential codes were necessary to unlock theatre nuclear weapons based in Europe (through a series of message authentication steps based on sealed code books)[73] and therefore without presidential authorisation the weapons could not be used. The President could thus decide simply to withold the codes in order to prevent any US nuclear weapons being used in the theatre.

The selectivity of Category D and F PALs (which were retrofitted to Lance, artillery shells, gravity bombs and other, but not all, theatre nuclear forces) also meant that high-level selectivity could determine just which weapons were released and in what number. The President could release only nuclear artillery shells for example if he wished to prevent longer range systems being used. It was equally possible, of course, that the President could provide blanket release of all US nuclear weapons in the theatre.

To understand how command and control of these weapons was exercised in NATO it is useful to take the discussion through a step-by-step account of the means by which weapons would have been taken from peacetime deployment, made ready for use and used.

All US nuclear weapons in the NATO theatre in peacetime were under the direct custody and control of the United States. Their whereabouts, condition, readiness and security were all subject to US control and the 'two-person' rule. WWMCCS centres in Europe stored and updated information about the weapons on both the Nuclear Capability Plan (NCP) and the Nuclear Weapons Accounting System (NWAS) databases.[74]

Theatre weapons, except those on naval vessels, were stored at one of six different types of facility:

1 Type A sites stored surface-to-surface missile warheads and nuclear projectiles (prior to their removal from Europe they also stored Atomic Demolition Munitions – ADMs). They also stored Prescribed Nuclear Loads (PNLs) which was the term for a preset quantity of weapons earmarked for a particular delivery system in the field. Once PNLs and other weapons have been deployed for war Type A sites are anticipated to be closed or abandoned;
2 Type C sites stored nuclear gravity bombs for NATO strike aircraft. They were usually co-located or close to Type Q sites;
3 Type Q sites were those at which QRA aircraft were deployed ready for prompt take-off. They provided custody of QRA PNLs and a take-off area for the aircraft. They were close to or co-located with Type C sites to enable rapid back-up or reloading of QRA and other NATO strike aircraft;
4 Type F sites provided support for Nike-Hercules air-defence missile units. The warheads, mated to missiles, were kept in storage buildings in a launching area to enable rapid operation;
5 Type S sites provided maintenance, repair and reporting services for Type A, F, and Q sites. They also stored PNLs and Prescribed

Nuclear Stockages (PNS – stockpile weapons to support operations but not earmarked for initial PNLs) weapons;

6 Type Z sites provided rear area storage for air-defence and ground nuclear weapons in the SACEUR reserve (that is weapons not earmarked for initial combat use in the event of conflict). They also stored PNL and PNS weapons.[75]

A typical storage site was protected by electronic systems and military personnel and used hardened controlled environment storage facilities (termed igloos) to contain the nuclear weapons.

The A, C, S and Z sites differed mainly in their size. Type A sites, the smallest, typically consisted of no more than two to four igloos; Type C and Type S of usually 9 to 14 igloos; and Type Z of between 18 and 22 igloos. The number of weapons in an igloo was normally between 12 and 25.[76]

At their peak in the mid 1980s there were reportedly more than 100 storage sites based in seven European NATO countries.[77] Of these almost half were in Western Germany including two very large Type Z sites at Fischbach and Miesau the main storage facilities for the two US Army Corps deployed in Western Germany.[78]

The custody and control of these weapons was carried out by US 'special weapons' custodial units, based at Pirmasens in Western Germany, estimated to number between 10000 and 25000 personnel in Europe.[79] In peacetime these units were under US lines of command, responsible to CINCEUR, and managed the weapons with the ageing European Command and Control Console System (ECCCS).[80] ECCCS was controlled from Pirmasens and there were network switching centres at Mount Parnis in Greece, Sabin Tepesi in Turkey and Pruem in Western Germany and user terminals at all Nuclear Weapons Storage sites (NWS).[81]

In a sufficiently serious crisis period in which the use of nuclear weapons was contemplated steps would probably have been taken to remove the weapons from storage and deploy them at operational sites to prepare them for use. The main reasons for this were, firstly, that in storage the weapons made a lucrative target and, secondly, that they could be needed rapidly for use. One source indicates that some weapons were intended to be taken from storage and deployed when NATO raised its alert level to LERTCON 2.[82] Although this is the point at which forces would 'Chop' from national to NATO command, US nuclear custodial units would remain under US control.

At some point during the alerting process custodial units would probably have switched from using the peacetime ECCCS to the HF radio Cemetery Net system.[83] Cemetery Net was installed in Europe between 1976 and 1982 primarily to provide the means by which the US could transmit the EAM to nuclear forces at storage sites, delivery units and command headquarters.[84] There were Cemetery Net switching centres at sites at Incirlik in Turkey and Edingen in Western Germany and user terminals at all storage facilities and mobile terminals for field sites.

Cemetery Net, however, was criticised for being insecure and subject to jamming. It was replaced in the second half of the 1980s by Regency Net, another HF network, which provided:

> HF, secure . . . voice and record communications; network [linking] command elements and nuclear storage detachments and teams; custody and control in peacetime;[85] and release and employment in wartime.[86]

Regency Net, which became operational in 1988, provided secure and survivable voice and data communications from three types of terminals. The first was a 'shelterised force terminal' used for ground locations (such as storage sites) or vehicles (such as warhead or weapon transporters). The second was a 'team terminal' for mounting on jeeps and on similar European host nation vehicles (these also include back-pack terminal). The third was an unshelterised version of the force terminal (its distinction from the first type is unclear).[87]

Regency Net could be used to order US nuclear weapons from storage to the battlefield and also to authorise their release in the field. It is possible that weapons taken from storage would have been released at the same time. In view of the hostility (both electronic and explosive) of the battlefield release could have been considered necessary to hedge against a subsequent inability to maintain connectivity with the weapons once they have been deployed closer to the battle area. It is also possible that weapons would have been dispersed without being released relying on the improved capabilities of Regency Net to enable the EAM to reach the weapons when required.

After dispersal some weapons would be taken to Special Ammunition Supply Points (SASPs) for prompt use while others would be taken to Field Storage Locations (FSL) for security. The number of SASPs and FSL far exceed the numbers of fixed storage sites thus far discussed. In terms of command and control there would be inevitable problems retaining connectivity with a large number of field deployments both near and rearward of the battle area.

To consider how the weapons would be used we need briefly to discuss the US and NATO command structures. The use of nuclear weapons required that the two command chains and the forces they control were co-ordinated into a whole (rather like two halves of a walnut). One half of the command chain was the NATO command structure which extended from the DPC/NAC through SACEUR to the Corps level and on to the field units. The second was the US national European command structure which controlled both the custodial units with possession of US nuclear weapons and the PAL teams.

In the case of a US military unit with control of, for example, a nuclear artillery gun the unit and its weapons would have passed under NATO command during the chop. The custodial unit, however, remained under US national control. For nuclear artillery shells to be fired necessitated that custodial units with shells met up in or near the battlefield with NATO forces possessing the correct delivery system. To fire a shell the delivery unit personnel must be in receipt of a valid firing order while the custodial unit must be in receipt of a valid EAM. Thus there were two chains of command and two communications systems operating side by side. Delivery system units relied on the tactical communications systems discussed in the previous chapter while those of custodial units use Regency Net.

In the case of US forces under NATO command in this scenario there was also the possibility that the forces would be responsive to national rather than NATO control and that the US could execute nuclear strikes by ordering both custodial units/PAL teams and national forces to use the weapons. This implies, however, that the views and advice of NATO allies would have been ignored or would not have been sought.

The exact steps that would be taken to seek and obtain the authority to use nuclear weapons would in the end very much depend on circumstance. The request for weapons release from a field commander would be passed up through the command chain to NATO's political bodies the NAC or DPC and requests would then be made to the nuclear powers (the US and UK) for the release of the requested weapons. If this was forthcoming there would, at least in theory, follow a discussion within the NAC or DPC to ensure that the views of at least those NATO members most involved (see Chapter One) were heard. Once agreed the order would then be passed down the command chain back to the Corps commander authorising the use of a Selective Employment Package (see Chapter Two). Simultaneously the nuclear power would authorise the release of nuclear forces through custodial units. The US Army manual FM100-5 which deals with nuclear weapons release indicates that this process could take as long as 24 hours.[88] In command and control exercises such as Wintex this process has reportedly taken as long as sixty hours,[89] by which time the forces that made the original request may have been overrun or defeated. This analysis raised doubts about the slow and cumbersome nature of the release procedures.[90]

There were however a number of reasons to cast doubt on these criticisms. Firstly, the 'time and circumstance permitting' phrase in the Athens guidelines suggested that the nuclear powers need not consult with their NATO allies if to do so would be likely to delay a nuclear response which was deemed necessary.

Secondly, it was not necessary for the request for the use of nuclear weapons to come from a Corps commander. Authorities at the top of the command chain could authorise a 'top-down' rather than a 'bottom-up' release.[91]

Thirdly, even if a Corps commander did initiate the request for nuclear weapons he need not have to wait for imminent defeat before doing so. He could request the use of nuclear weapons from his senior commanders well in advance, perhaps even in advance of actual combat. General Bernard Rogers, former SACEUR, testified before a US Senate Committee in 1983 that:

> Prior to [SACEUR's nuclear release request] there would have been a warning message that I was probably coming to ask for release, And even prior to that, in order to get the political authorities thinking in terms of giving this permission, I would have sent what I would call an early notification message to them.[92]

Finally, SACEUR need not pass the release orders down through the command chain. In the 1980s SACEUR deployed the SACEUR Control Alerting and Reporting System (SCARS II) which was installed at all major headquarters down to PSC and used both the IVSN and TARE networks as a bearer (see Chapter One).[93] SCARS II enabled SACEUR to directly issue release orders to relatively low command levels.[94]

These caveats indicate that the nuclear powers in NATO could have recourse to nuclear weapons far more quickly than FM100-5 indicates and without necessarily consulting NATO allies.

COMMAND AND CONTROL OF BILATERAL THEATRE NUCLEAR FORCES

The command and control of NATO bilateral forces differs only marginally from the discussion in the previous section. For bilateral forces to be used required that the forces of the European ally having operational control of nuclear weapons delivery systems (artillery guns, Lance launchers, and so forth) met in the field with US custodial units having possession of the nuclear weapons (or warheads).

Providing that both were in possession of valid orders and authority the two elements of the weapons system could be mated, the weapon armed and used. Parallel command chains complicated this process, and in the case of bilateral forces the process was complicated further by differing communications systems, different languages, and differing procedures.

The European forces with operational control of delivery systems would normally be under NATO control sometime before receiving nuclear weapons from custodial units. In theory at least the weapons would therefore remain in NATO control.

However, once the military forces of the non-nuclear allies (ie possessing no national nuclear weapons) became nuclear at the point when weapons were handed over by the American custodial units the relevant European allies would have a strong incentive to maintain a high degree of national control over their own forces to ensure that the nuclear weapons in their possession were not used without clear national authority.

Each of the European allies deployed robust survivable strategic communications to ensure connectivity between senior political and military leaders and national forces in the field. For the European allies with bilateral nuclear force arrangements these strategic systems were:

Belgium	BEMILCOM
Greece	HEDICS
Italy	Tri-Service Network
Netherlands	ASCON
Turkey	TAFICS
United Kingdom	BOXER/UNITER
Western Germany	GAFDIN/GAFACS[95]

The US equivalents of these, to provide a comparison, would be the AUTOVON/AUTOSECVON/ETS/DSN networks.

At and below the Corps level each of the European allies also, as we have seen, had national tactical communications. Although prior to the use of theatre nuclear weapons forces would have been passed to NATO command following the chop, they would still be in constant and direct communications with national military and political leaders. The means therefore existed for national command and control of the forces and any national orders contrary to those of NATO were likely to be obeyed by national forces (a discussion of this and related points in the specific case of the United Kingdom's bilateral nuclear weapons is included in the next chapter).

It was also possible that once armed nuclear weapons were handed over to European allies for use in NATO SEPs, those weapons could also be directed to national purposes through national lines of command and control.

CONCLUSIONS

In the United States the President had primacy in the control of nuclear weapons though his authority was exercised through the National Command Authority (NCA) which comprised at least the President and the Secretary of Defence and usually also the Chairman of the Joint Chiefs of Staff (CJCS).

The authority of the President was carried out through release and authentication codes which amounted to direct and physical political control of most of the US's nuclear weapons in the theatre (the exceptions were naval nuclear weapons which were not subject to PAL or related technology).

Although the power of the President was in theory absolute in practice it was subject to the joint decision-making of the NCA, to an Anglo-US commitment to consultation prior to the use of nuclear weapons by either party in any theatre, and within NATO to the commitment to consult with NATO allies. None of these commitments were, however, mandatory and the President was ultimately free to order the use of nuclear weapons or to delegate his nuclear authority and release codes virtually without limit.

The United States deployed a vast command and control system to control the global deployments of its nuclear forces. At the strategic level the most robust and assured element of this system – the MEECN – overlapped in specific mission areas with US nuclear forces assigned to NATO in the European theatre.

At the theatre level a multiplicity of systems were deployed both to link the continental US to the European theatre and to control US nuclear (and all other military) operations within the theatre. In common with NATO command and control systems the latter were introduced in an evolutionary way but in contrast to NATO these systems – such as the ECCCS, ETS, and AUTOVON/AUTOSECVON/AUTODIN – were designed for theatre-wide operations. US national connectivity across the theatre was substantially better than that of NATO's multinational forces.

US nuclear weapons within the theatre were controlled by robust and dedicated command and control systems and political control was enforced through the widespread employment of PALs (designed initially for exactly the purpose of controlling US nuclear weapons deployed outside the continental USA).

As NATO's principal nuclear armourer, yet bound by its own constitutional limitations, the US had with NATO developed a double and parallel command chain designed to facilitate nuclear-sharing yet impose and ensure American control of nuclear warheads until almost the moment of firing. This arrangement also facilitated autonomous US nuclear operations across the theatre if required.

The European bilateral nuclear forces had also introduced robust command and control systems to support their respective elements of nuclear operations. In effect these arrangements gave the relevant European nations a limited degree of *de facto* control over those nuclear weapons earmarked for their use. Until the moment the US nuclear munitions were handed over this in practice amounted only to the negative control of a veto. Once full operational control of the relevant weapons was handed to the European bilateral forces the initiation of nuclear war by the respective European nations became possible.

4 Command and Control of British Nuclear Weapons

INTRODUCTION

This chapter examines the command and control arrangements for British nuclear weapons. It considers both the command and control of UK national nuclear weapons and those of NATO which were planned for use by British armed forces under bilateral arrangements.

The chapter is organised into three sections. The first deals with political authority over British nuclear weapons and considers political decision-making both outside and inside NATO. The second describes the command and control of national strategic forces. The third details the command and control of national tactical nuclear forces.

The chapter is intended to widen the discussion of nuclear command and control in NATO by serving as a case study both of a national nuclear force integrated into NATO and as an example of one of the European participants in NATO bilateral nuclear-sharing arrangements.

POLITICAL AUTHORITY

Political authority over the use of British nuclear weapons reflected both the national character of the force and the commitment of that force to the NATO alliance. Successive governments pledged British nuclear weapons to the NATO alliance but always made it clear that circumstances existed in which the United Kingdom could use its nuclear weapons independently.

The commitment of British nuclear weapons to NATO had its genesis in the December 1962 meeting in Nassau between US President Kennedy and British Prime Minister Macmillan. The final communique of that meeting spelled out the commitment of British forces to NATO yet made explicit the possibility of independent national use:

> . . . except where Her Majesty's Government may decide that supreme national interests are at stake, these British [nuclear] forces

> will be used for the purposes of international defence of the Western alliance in all circumstances.[1]

The stated position of past British governments was that any use of British nuclear weapons would be subject to British political processes and that the final decision about their use would rest '. . . solely with the British Prime Minister'.[2] Within the NATO context, however, the United Kingdom was committed, if not bound, to consultation within the alliance prior to any use of British nuclear weapons. It is therefore important in any examination of British political authority over the use of nuclear weapons to consider the decision-making process both outside and inside the NATO framework.

To deal first with decision-making outside NATO, that is the national process by which the decision to use British nuclear weapons would be taken and carried out. Such a decision would have depended upon the political and military context at the time.

If the decision to use nuclear weapons were taken without hinderance it could be argued that it would have been taken by the Prime Minister, advised by full Cabinet and ideally with the agreement of Parliament (which constitutes the ultimate political authority in the country).

It is perhaps more realistic to suggest that such a decision would have been taken during a period of crisis or even conflict. Historically, crisis or wartime decision-making has involved the *ad hoc* formation of 'war cabinets' usually involving fewer individuals than full cabinet in order to facilitate swift decision-making and protect security. A recent war cabinet was that set up to direct the Falklands campaign in 1982. The political make-up of this body included the Prime Minister, the Deputy Prime Minister, the Foreign Secretary and the Defence Secretary.[3]

It is also possible that the national import of any decision to use nuclear weapons might have persuaded a government to include some senior parliamentary opposition figures in the decision-making process and there are many precedents for this suggestion.

For as long as there was no immediate threat to the United Kingdom decision-making was likely to be carried out in 10 Downing Street, the Prime Minister's London residence. However, if a threat arose a number of alternative command centres were available from which the Prime Minister and war cabinet could operate, all of which were underground.

Beneath the Ministry of Defence buildings in Whitehall was a primary command centre code-named Pindar which was connected by underground tunnels to Downing Street.[4] This centre was completed in the latter part of the 1980s and replaced the former Cabinet Office control centre known at Cobra.[5] In the event that Pindar became insecure, perhaps because of a threat to the capital, political leaders could be 'evacuated and established in protected headquarters far from London'.[6] What is not clear is exactly when political leaders would have evacuated Downing Street for Pindar, and Pindar for the countryside headquarters.

The primary alternative headquarters outside London was the vast underground complex at Hawthorn near Bath.[7] According to press reports this complex was over 30 metres beneath the ground and was some 54 acres or two million square feet in area. It was understood to be entered at some points by escalators similar to those of an underground railway system and a small rail-type network of transport vehicles was thought to be in operation underground. The centre is understood to be have been provided with all the necessary amenities to sustain occupation and to be connected to Britain's nuclear weapons infrastructure by a variety of communications systems (see below).

The Hawthorn complex was itself part of a larger group of command centres which extended several miles beneath the neighbouring town of Corsham in Wiltshire.[8] Other parts of the overall complex included the headquarters of the Defence Communications Network (DCN – see below) to the north of Hawthorn and the headquarters of the UK Land Forces Army Command to the east.[9]

No part of the complex was able to survive direct nuclear attack. Because of this it has been suggested that smaller publicly unknown centres probably existed elsewhere in the United Kingdom for specific individuals such as the Prime Minister or the Royal family. Such centres could escape nuclear targeting by virtue of the secrecy of their location.[10]

The vulnerability of the British Prime Minister at any of these sites questioned the wisdom of concentrating the authority to use nuclear weapons in the hand of a single individual at the top of an hierarchical command system for reasons which have already been considered with respect to the American President (see Chapter Three). In reality the decision to use British nuclear weapons did not depend on the availability and decision of the Prime Minister alone. The evidence suggests rather that the United Kingdom had developed a nuclear weapons command and control system which provided primacy for the Prime Minister in decision-making, but which could continue to function in his or her absence.

While the Prime Minister had the sole authority over the use of British nuclear weapons he or she did not in fact take the decision alone. The United Kingdom was known to operate a 'two-person' rule throughout the nuclear command chain from the Prime Minister to the serviceman in the field.[11] The Prime Minister was not therefore able to order the use of nuclear weapons alone, but could do only so with the assistance of at least one other person, presumed in most instances to be the United Kingdom's most senior military figure, the Chief of the Defence Staff (CDS).

The Prime Minister was not thought to hold the codes which were necessary to order the use of nuclear weapons.[12] At least one source indicates that these were kept with '. . . a civil servant and a military officer'.[13] These codes were intended to be transmitted to the appropriate military headquarters – Northwood for the Royal Navy and High Wycombe for the Royal Air Force – and checked there against codes already in the military's possession. A match of the two would indicate that the source of the instruction was, in fact, the Prime Minister and that therefore the order was valid. The respective commanders would then have, with their own systems of coded instructions, ordered and executed the use of nuclear weapons.

The Prime Minister did not, therefore, seem to exert any direct physical control over the use of nuclear weapons beyond the authority of office. In peacetime or crisis this was obviously enough to give the Prime Minister the principal role in decision-making.

In wartime circumstances, when the survival of the Prime Minister could be in question, plans may have existed – or may have hastily been prepared – for the delegation of command authority from the Prime Minister. The Prime Minister was able to delegate the authority to use nuclear weapons in advance to specific individuals or delegation may have occurred by virtue of his or her death or incapacitation. Either way some provision probably existed for the continuity of high level control.

One possible approach was the preparation for the transfer of authority to subordinate political figures. In the US a list of Presidential succession was drawn up which detailed the senior political figures who would assume presidential powers in the event of the President's death or incapacitation (see Chapter Three).[14] No evidence has been found that a similar succession list existed for the United Kingdom.

However, following the Irish Republican Army (IRA) attack on British cabinet members at the Grand Hotel in Brighton in October

1984, the constitutional repercussions of the possible loss of the Prime Minister and other cabinet members were addressed unofficially by the Attorney General Sir Michael Havers. His assessment was that temporary control would have been assumed by the Deputy Prime Minister until such time as a new government could be formed.[15] This implied an implicit delegation of the Prime Minister's role but did not indicate the extent or nature of such delegation, nor of course whether such delegation transfered nuclear command authority. However, political delegation of command authority in a period of crisis or war was a prudent measure to ensure continuity of political control. The likely circumstances of such a transfer of power must, nevertheless, cast doubt on this type of arrangement for the United Kingdom.

It is probable that during a crisis some, if not all, of the senior political leaders would have been with the Prime Minister as part of the war cabinet. Their fate consequently would be the same as that of the Prime Minister and the point of any delegation would therefore be moot. Similarly, there were only a few alternative command headquarters available and even if senior political leaders were sent to these centres it was unlikely that they would escape targeting or, if they did, that they could function as national leaders from one of the less well equipped bunkers.

It was therefore possible that delegation of the authority to use nuclear weapons would pass directly to military leaders. Senior military leaders would have been operating from underground headquarters and these may themselves have been targeted. It can consequently be argued that the survival chances of these individuals was perhaps no greater than that of the Prime Minister and cabinet. As a result delegated authority could have passed still further down the command chain. To understand how this would be effected it is important to pause and consider exactly how the control of nuclear weapons was exercised in the UK.

As already pointed out the Prime Minister exerted no direct physical control over nuclear weapons. The US made Britain aware of electronic PAL technology in the 1960s which would have enabled the Prime Minister (or other political leaders) to enforce physical control of nuclear weapons.[16] The UK, however, rejected this technology preferring instead to rely on organisational means to control the weapons.[17] Obedience to political authority was enforced by complex procedural arrangements which relied on the co-operation of numerous individuals in the command chain and on the use of authenticating codes to establish the validity of orders received. The absence of physical controls meant that those who had possession of nuclear weapons (for example a submarine commander and his officers) have the *de facto* ability to use those weapons even if they did not have the authority to do so.

This arrangement was thought to have important virtues for deterrence purposes. In peacetime the arrangements were thought to be adequate for the control of weapons to ensure there was no unauthorised use. In crisis or wartime, and this was the crucial point, the ability to retaliate with nuclear weapons was already in operational military hands in case the political and military leadership was destroyed. Delegation of command authority thus passed directly to nuclear weapons field (i.e. operational) commanders if political and military leaders was destroyed. In only one case, that of the Polaris submarine, has this position been clearly articulated. According to one author, the Polaris submarine commander was (with a second officer to adhere to the two-man rule) '. . . capable of launching [Polaris] missiles of [his] own volition'.[18] He was authorised to do so only if he believed '. . . that Britain [had] been destroyed by a nuclear attack and no responsible person [was] in a position to transmit orders [to the submarine]'.[19]

The purpose of this arrangement was thus to attempt to ensure retaliation and thereby underwrite Britain's deterrent posture. The risks of accidental, unauthorised or unintentional use of nuclear weapons – risks which prompted the US to adopt physical control measures – seemed, however, largely unaddressed by this approach. Perhaps even more important was the question of whether organisational means alone could enforce political control of nuclear weapons during a conflict.

As has previously been stated, the nuclear weapons of the UK were committed, with certain reservations, to the NATO alliance. As a result of this, in addition to the foregoing, the UK was committed to consultation within NATO should it wish to use nuclear weapons, or make its nuclear weapons available for use, in the NATO context. As was described in Chapter One, the UK was subject to the consultative mechanisms and political guidelines for the use of nuclear weapons set out by the North Atlantic Council.

Had the United Kingdom chosen to use nuclear weapons within NATO she was obliged to consult with other alliance members and await their decision before weighing their views and arriving at a final decision. Had, conversely, a NATO commander requested the use of British nuclear weapons, all NATO governments would have been informed of the request and the UK would once again await the views of her alliance partners before making a decision.

In practice time pressure and force of circumstance could have precluded this elaborate consultation and permitted only brief discussions or reliance on preplanned decision-making. The UK, like the USA, was not bound to consultation with NATO allies if time and circumstance did not permit. The UK undoubtedly had the means to '. . . nuclearize [the] situation in central Europe' independently of the rest of the alliance.[20] The UK could also have, by the same token, refused to commit her nuclear weapons to NATO.

The United Kingdom was committed to one other process, outside the NATO framework, of consultation on the decision to use British nuclear weapons. This was the 1952 bilateral Anglo-American agreement discussed in detail in chapter Three which committed both parties to consultation prior to the use of nuclear weapons by either party provided that time and circumstances allowed.

STRATEGIC NUCLEAR COMMAND AND CONTROL

Command of the United Kingdom's strategic nuclear forces – the Polaris submarine fleet – was exercised from the Royal Naval headquarters at Northwood in north London. Northwood was a joint UK and NATO facility housing both British and NATO staffs. The commanding officer at Northwood (invariably British) held both the national rank of the Royal Navy's Commander-in-Chief Fleet, and the two combined NATO ranks of Allied Commander-in-Chief Channel (CINCHAN) and Commander-in-Chief Eastern Atlantic (CINCEASTLANT). The commander was served by two separate staffs – one British, one NATO – operating from two separate underground operations rooms at Northwood.[21]

Orders to use Britain's strategic nuclear forces would have been passed to Northwood from the Prime Minister together with the authenticating codes. A number of communication systems were available for this including the Defence Communications Network (DCN) which, *inter alia*, was reported to '. . . link central government headquarters and military bases by telex and telephone'.[22]

The DCN was in fact the blanket term given to a network of HF radio, SATCOM, commercial cable and microwave/troposcatter communication systems which linked the UK to her global military assets. The DCN was divided into two principal elements. The first linked UK command authority with British forces in Western Germany. The second handled communications to all other British military assets around the world. The DCN systems acted as the carrier for the Defence Secure Speech System (DSSS) for secure national and global voice communications.[23]

Within the UK the DCN was supplemented by a microwave network of hardened centres known as BACKBONE. From BACKBONE a network of commercial microwave stations linked the command centres and principal military bases.[24] The Royal Navy also operated a hardened (and publicly unnamed) telephone system based on fibre optics which may have linked Northwood and the Prime Minister.[25]

Under the Pindar project umbrella, of which the Pindar command centre beneath Whitehall (described above) was part, the UK developed a national command, control and communications network for crisis management based on the concept of dispersing command and control functions along ARPANET/LAN principles (see Chapter Three). This system was not fully implemented by 1990 but completed components used the DCN and associated hardened bearers to carry telephone and computer data.[26]

Despite the preparations for assuring communications connectivity Northwood itself was vulnerable to attack and some preparations may have existed, in case Northwood was attacked or destroyed, for the assumption of operations at other sites, perhaps at the principal subordinate command centres at Plymouth and Pitreavie Castle. These, however, were also vulnerable and could not have been expected to survive any longer than Northwood.

The exact means by which orders would be passed to the Polaris fleet from Northwood was dependent upon circumstances and therefore it is necessary to review the systems upon which the United Kingdom relied for attack warning since these determined the threat perception and therefore, indirectly, how the British response was intended to be carried out.

Early indications of imminent nuclear attack against the UK would have come from US intelligence sources, either directly as part of the close Anglo-US intelligence relationship or through a pooled assessment of threat made available to NATO leaders. The UK had no satellite intelligence gathering capability – a vulnerability the Zircon satellite was intended to address [27] – although she still had many global ground-based intelligence assests which had some role in acquiring relevant intelligence.[28] The first indication that a ballistic missile attack had actually begun would probably have reached the UK from the US

NORAD headquarters inside Cheyenne Mountain in Colorado. Information from the DSP satellites (see Chapter Two) would have been passed to a variety of US and allied command centres including Northwood and the RAF headquarters at High Wycombe.[29] The DSP, however, was an ageing system and its capabilities were in question.[30] The importance of this for the UK was whether NORAD could have determined if the missiles are heading for UK or the USA (or both) at this early stage of attack. If NORAD was unable to determine the intended target nation then the USA would have considered itself under attack. In such a situation the UK could not have been assured of priority in the NORAD alerting process.

Confirmation of initial attack warning would probably have occurred within a few minutes from the ground-based Ballistic Missile Early Warning System (BMEWS) radars at Flyingdales. Flyingdales housed three AN/FPS-49 scanning and tracking radars with a range of over 3000 kilometres.[31] These provided surveillance over most of Eastern Europe and deep inside the Soviet Union (see Chapter Two).

Fylingdales was an anomaly of Anglo-US cooperation. Its principal function was as one of three BMEWS radars (the others are at Thule In Greenland and Clear in Alaska) which provided NORAD with warning of attack on the US. To this end the radars, computers and communications (to the US) were all paid for by the US Air Force. Under the UK–US construction agreement of 1960 the UK was responsible for the site, the power and the housing at Fylingdales.[32] The RAF also staffed the operations room at Fylingdales, giving it effective control of operations[33], and had installed its own communications (with US agreement) so that attack warning could be provided directly to British command centres at Whitehall, High Wycombe and Northwood.[34] The communications systems which linked these centres were those already described (DCN, BACKBONE, Pindar) together with the service-specific fibre optic systems which were introduced later (see below).

Fylingdales AN/FPS-49 radars were capable of tracking ballistic missiles trajectories and hence could determine whether an attack was destined for the USA or for Europe and the United Kingdom. There was however a questionmark against the length of warning time Fylingdales could provide for the UK. Theoretic flight times of Soviet ballistic missiles to the UK varied up to approximately 15 minutes depending on their origin and trajectories.[35] At least one source indicated that BMEWS radars would not be able to confirm attack until ten minutes after launch, both because it was necessary for rising missiles to appear above the curvature of the earth before they could be detected and because there was a delay while Fylingdales verified its initial sighting.[36]

For the United Kingdom such a delay could have meant that no warning of attack was received, or at best that only around five minutes warning time was available. One British government publication states that '. . . no particular warning time can be guaranteed', and went on to say that not less than three minutes warning time was anticipated.[37] This refers to the warning time Fylingdales was able to provide for the civilian civil defence organisations and may therefore have been a slight underestimate.

The key point is that the BMEWS radars were designed to provide early warning for the USA. When they were constructed in the early 1960s the radars addressed the threat from Soviet land-based missiles which had flight times to the USA of approximately 30 minutes. The radars were designed to provide warning within ten minutes of missile launch, a time-frame which would still have given the US 20 minutes in which to respond. No such margin of warning time was available to the UK. Moreover, being of 1960s vintage there were doubts about the ability of the AN/FPS-49 radars to track the numbers and reduced size of modern multiple warhead systems. One US source has indeed indicated that as early as the late 1970s BMEWS radars were 'unable to deal with the current standard of missile technology'.[38]

The limitations of Fylingdales were further underlined when addressing the threat of submarine-launched ballistic missile attack. Fylingdales was built to address a land based missile threat and as such its area of surveillance was orientated towards the Eastern Soviet land mass. There were, however, large areas of the ocean around the UK which were not covered by the radars at all. The vulnerability in this area was heightened further by the limitations in coverage of the area of the DSP satellites. Taken together these gaps indicate clearly that a ballistic missile attack on the United Kingdom from the Northern Atlantic against the UK could have gone completely undetected.

In response to the age and unsuitability of the BMEWS radars and in response to the gaps in UK early warning the BMEWS radars were upgraded to phased-array systems in the late 1980s which was intended to provide far quicker and more accurate target acquisition and tracking capabilities. For Fylingdales this meant a complete upgrade of computer and radar technology. The new radar was designed to provide 360 degree coverage around the UK and to be capable of discerning smaller and more numerous objects to provide better and more timely attack assessment.[39]

Against air-breathing missiles and aircraft, the United Kingdom was less reliant on the US. The UK developed through the 1980s an extensive and modern air defence network known as the UKADGE (United Kingdom Air Defence Ground Environment) which was integrated with that of NATO and France. This network itself superceded a complex air defence network, elements of which reached back to World War II.

The UKADGE network was based on 12 transportable General Electric, Marconi Martello, and Plessey/ITT 3-D radars deployed at eight sites around the UK.[40] Each radar had a range of approximately 300 miles. The mobility of the radars allowed them to be towed to presurveyed sites away from targeted area (such as known RAF bases) to avoid being destroyed. Decoys could also be used to lure homing missiles away from the radar emmissions.[41]

In addition to the ground-based radars UKADGE also intended to utilise airborne early warning aircraft (AWAC). After a series of technical shortfalls and cost overruns in the 1980s the UK abandoned a British AWAC system based on the Nimrod aircraft and opted to purchase US Boeing AWACS aircaft. Delivery of these aircraft had not begun by 1990 and therefore for the period under consideration the UK relied on ageing Shackleton aircraft with limited operating capabilities.[42]

The ground and air-based air defence radars were intended eventually be linked to the Uniter communications system, a hardened and protected fibre optic land-line and microwave network. Uniter carried early warning information to the Air Defence Operations Centre (ADOC) at RAF High Wycombe, to the alternative ADOC at RAF Bentley Priory and to a number of other air defence command centres. At these sites air defence information was computer processed into a real-time picture of the air environment around the UK (this system also used information from the civil and military radars in the civilian National Air Traffic Control System – NATS). This overall picture of the UK air defence environment could then be integrated with that of France (STRIDA) and NATO (NADGE) to provide a comprehensive picture of the air-breathing threats to NATO.[43]

One flaw in the UKADGE system resulted from the failure of initial planning in the mid 1970s to take account of the threat of cruise missiles.[44] Low-flying cruise missiles, particularly if launched from submarines could have flown below the scanning arcs of the ground-based radars and thereby avoided detection. Although this vulnerability was to some extent addressed by the AWACS which have a 'look-down' capability, it spurred the development of Over-The-Horizon (OTH) radar technology in joint US–UK ventures. At least three OTH radar systems were developed in the UK. The first, a joint US–UK project was the subject of the COLD WITNESS project and a prototype radar using skyward propagation was built at Cricklade near Swindon.[45] The second, a British only venture, using surface propagation was established at Angle in Pembroke near Wales.[46] A third project, a joint UK-US Navy operational system was due to be deployed at St David's airfield in Pembroke with the reception antenna at Cricklade, but was incomplete when flexible response was abandoned.[47]

To communicate with the Polaris fleet the United Kingdom relied on three VLF transmitters at Criggion, Rugby and Anthorn. Criggion and Rugby were civil as well as Royal Navy establishments while the transmitter at Anthorn was actually operated by NATO.[48] These transmitters had aerials consisting of approximately two miles of above-ground cable and a power output of 60–150 KW.[49]

In wartime if the security of these transmitters came under threat the UK had a variety of back-up communication systems to maintain contact with the Polaris fleet. These reportedly included a number of mobile transmitters which could exploit the radiation potential of structures such as bridges and railway lines for use as antennae.[50] These mobile transmitters could be dispersed in crisis or wartime.

There were also a number of Low Frequency (LF) transmitters available. Like VLF these transmitters allowed Polaris to receive signals while still submerged, though only to a lesser depth of perhaps 3–4 metres.[51] The Royal Navy operated a number of LF transmitters, the most important of which were those at Inskip and Crimond.[52] In addition the Royal Navy had plans to use the British Broadcasting Corporation LF transmitters (which usually transmitted BBC Radio 4) at Droitwich in an emergency.[53]

Finally, there was some ambiguity about the possible use in Polaris fleet communications of US TACAMO aircraft (see Chapter Three). Both the USA and France used airborne VLF communications for their ballistic missile submarine fleet while Britain decided not to develop such a capability (or seek to acquire it from the USA). A single report in the defence press suggested that Polaris had made use of the US TACAMO aircraft to pass on signals from Northwood.[54] Even if this was untrue for peacetime operations it raised the possibility of US TACAMO communications with the Polaris fleet in wartime.

Polaris could also have used the Royal Navy's standard ship-to-shore communications systems (including MF, HF, VHF, and UHF) but these required the submarine to raise an antenna above the water and risked compromising the vessel.

During the 1980s the VLF transmitters were upgraded with better computer technology and transmitting equipment which improved the quality of transmissions. During the 1980s the existing range of transmission media was supplemented by work on an extremely low frequency (ELF) transmitter at Glen Garry in Scotland. ELF transmissions were able to penetrate water to a greater depth (up to 120-180 metres) than VLF and could extend the patrolling speeds at which submarines were able to receive transmissions. ELF transmissions were, however, extremely slow. A simple three letter transmission could take several minutes to send while a longer message – such as the instructions for nuclear retaliation – '. . . would take hours and [would] therefore not be practical'.[55] Furthermore ELF transmitters had to be even larger than those for VLF signals because of the huge wavelength of the transmissions (the US ELF site in Michigan state, for example, has 56 miles of above ground cable for an antenna[56]). This meant that the vulnerability of VLF transmitters to sabotage, conventional or nuclear attack was not addressed by the ELF approach. After almost ten years of exploratory work the UK cancelled its prototype ELF project at Glen Garry in Scotland with little explanation.[57]

Communication with the Polaris fleet was thought to be guaranteed by the large number of different communications systems available for the purpose. Speaking about Polaris communications in 1980, the then Deputy Under Secretary of State said: '. . . our main defence is not the invulnerability of any single (communication) system but the multiplicity of systems available to us – we have the freedom of redundancy'.[58] An examination of communications vulnerability indicates, however, that this may have been an optimistic claim.

Firstly, the principal media of submarine communications, VLF, was prone to difficulties. VLF penetrated only to a depth of 10-15 metres requiring the submarine to move closer to the surface for reception and also required the vessel to slow down.[59] The reception of VLF was also problematic. Polaris primarily relied on a trailing wire antenna which was towed behind the vessel at a distance of several thousand feet. The trailing wire was bi-directional and did not permit

360 degree reception from shore-based transmitters.[60] The submarine had to patrol at specific tangents to the shore transmitters or else it risked the degradation or temporary loss of transmission signal. Polaris could also use either floating buoy or floating wire antenna. The former involved floating a towed buoy to the surface some distance behind the submarine. This offered the prospect of utilising other radio frequencies (since the aerial floated to the surface). However, the buoys were prone to detatching, often sank when the vessel changed course (cutting reception), and could have given away the position of the vessel. The floating wire, as the name suggests, involved floating a long wire to the surface.[61] This also held out the prospect of utilising other frequencies but it too was prone to degradation if submerged and could also have given away the submarine's position.[62]

Secondly the ELF, VLF and LF transmitters were all large vulnerable soft targets for sabotage and conventional attack and there was little guarantee that they would be available in a conflict, particularly if an adversary were intent on putting them out of action.

Finally, the back-up mobile transmitters may have been better able to survive than the fixed VLF transmitters but they may still not have guaranteed assured communications with the Polaris fleet in wartime. The reduced power of the mobile transmitters together with the imperfections of the impromptu antenna were likely to result in weak and unclear signals, making reception uncertain.

Of course to assert that the vulnerability of these systems means that none would survive an attack is stretching the point too far. It is probably more realistic to conclude that some of the back-up systems might have survived but this would still not guarantee communications with the Polaris fleet.

The reasons for this assertion lay in the circumspect way in which the Polaris fleet was tied into the early warning system. Early warning of a nuclear attack was, as we have seen, passed to Northwood. To communicate attack warning and the orders to retaliate to the Polaris fleet Northwood would have passed on the information to the VLF transmitters and they in turn would send the message to the fleet. This could not have been achieved without some delay between the detection of attack at Fylingdales and the actual reception of the information on the Polaris submarines. Yet it has already been argued that nuclear missiles could have struck the UK with little or no warning. It was therefore possible that even if warning could be provided (and there is no guarantee that it could) both Northwood and the principal communications systems could have been destroyed before the message was passed to the submarine fleet. An enemy determined to isolate the Polaris fleet would also have destroyed the support command centres at Plymouth and Pitreavie Castle. Surviving back-up communications systems could therefore have been isolated from any meaningful political or military authority and their utility would be questionable.

Arrangements made by previous British governments support the contention that the Polaris fleet could have been isolated in a nuclear attack on the UK. Each took the step of authorising Polaris crews to fire nuclear weapons of their own volition under certain circumstances and each eschewed any physical control from land over the use of Polaris nuclear weapons. This was precisely because of the risk that after a nuclear attack on command centres and communication systems the Polaris fleet could be isolated from any meaningful political or military authority. Successive governments have reasoned that physical control over the use of nuclear weapons (such as PALs) or a requirement that submarine crews obtain political or military authority in all circumstances before firing nuclear weapons could, in the absence of guaranteed communications with the submarine, have risked the paralysis of the retaliatory nuclear force, something which could not be permitted.

A closer examination of the exact means by which Polaris submarines would have launched their nuclear weapons illustrates how the system was responsive to political or military authority if communications were maintained yet how the missiles could still be launched if the Polaris fleet was isolated.

While on patrol the Polaris submarine was dedicated to the defence of the NATO alliance. For this Polaris carried target tapes (which were fed into missile computers to direct the missiles onto targets) which were devised by the US Joint Strategic Target Planning Staff (JSTPS).[63] The targets British Polaris submarines were eventually allocated formed part of the European Nuclear Operations Plan (NOP) which in wartime would be directed by SACEUR.[64]

The Polaris submarine also served British national interests and for this it carried target tapes which were devised quite separately from NATO. The task of producing these tapes rested until the mid-1980s with the Navy and the Defence Intelligence Staff, but in early 1985 responsibility for British targeting was assumed by the Nuclear Policy Directorate at the Ministry of Defence.[65] This arrangement allowed Polaris to simultaneously meet NATO and national interests. Whether Polaris acted for NATO or the UK became a simple practical matter of selecting the appropriate target tapes in response to the appropriate instructions (which would have been provided in the event of a deteriorating international situation).

Under either circumstance, providing command authority on the mainland was intact, Polaris would have received the orders to launch missiles from Northwood. This would have initiated a series of procedures designed to observe the British 'two-person' rule and to hedge against collusion which might have resulted in the unauthorised firing of nuclear weapons.

The communications officer who received the coded signal from Northwood would have passed it to two officers – the weapons engineer and the first lieutenant – to check the codes against codes already in a safe on the submarine. This safe required two combinations to open it and each of the two officers knew only one combination. The codes in the safe would have matched those of the transmission from Northwood if the order was genuine. Having established the authenticity of the order a verbal instruction would be given by the Captain to bring the missiles to launch readiness. While the Captain remained in the control room the weapons officer would have removed a trigger devise (similar to the handle of a revolver) from another safe located in the Missile Control Centre (MCC) and connected it to the missile firing system. Once the missiles were readied for firing the submarine captain would have armed the firing mechanism by inserting a key (which is kept at all times round his neck) into the firing system and turning it. This would have illuminated a panel light in the Missile Control Centre and the weapons officer could then have fired the missiles in sequence with the trigger device.[66]

At each stage of authentication and firing at least two individuals were involved in the process and for the final stage the submarine captain and weapons officer were in separate rooms. Including the wireless officer, weapons officer, first lieutenant, captain and other crew members, around a dozen people had to be involved in the firing process. This was considered to provide a more than adequate check against the possibility of unauthorised use of the missiles.[67]

It can be seen that by merely omitting the authentication step the rest of the procedure could have been executed with complete autonomy by the submarine crew and the missiles launched. The launch autonomy of Polaris submarine crews was meant to safeguard the capability for nuclear retaliation even if the UK had been devastated by nuclear attack and the submarines isolated. There were, however, doubts that even this safeguard was guaranteed and it may be overly simplistic to assume that the isolation of the submarines would have provoked immediate retaliation.

If the Polaris fleet was isolated by a nuclear attack against the UK the overriding directive for submarine crews was not to retaliate but to monitor all forms of communication before making a retaliatory decision.[68] Retaliation would almost certainly have been postponed while efforts were made to ascertain the exact situation on the British mainland.

Many different circumstances could have obtained in the wake of a nuclear attack on the UK. For example a nuclear attack on the command and control infrastructure, designed to isolate the Polaris fleet, could have left many cities relatively unharmed. A retaliatory attack by Polaris may then only have provoked further attacks on the UK. Under these circumstances, Polaris crews could have reasoned that nuclear retaliation would be suicidal for the UK and could have held back for this reason. They may equally have been concerned that some surviving remnant of political or military authority in the UK wished to surrender rather than face national suicide. Such surviving authority could have been attempting to contact the submarines to order them to the surface or to return to port.

Whatever the scenario the key point is that the Polaris crews could have been uncertain of the situation and what was expected of them and might have postponed retaliation in the face of such uncertainty. As support for this kind of analysis one source quotes a Polaris commander as stating that '. . . you don't unleash Armageddon on the basis of radio silence'.[69]

In the NATO context VLF and other transmitters in the US or elsewhere in Europe could have been used to contact the British submarines with launch orders or orders to stand down (the significance of the report of the use by Polaris of the US TACAMO aircraft is apparent at this point). The use of Polaris missiles could have triggered a strategic nuclear war and hence the US would have had powerful motives to prevent such escalation if it was seeking to contain a war in Europe. If the US was already committed to a strategic nuclear conflict then the question of whether UK Polaris missiles were used or not would be of marginal relevance.

Either way there are powerful reasons to argue that past arrangements for the command and control of UK strategic nuclear forces did not necessarily guarantee nuclear retaliation.

TACTICAL NUCLEAR COMMAND AND CONTROL

British tactical nuclear weapons were almost totally committed to NATO. The stated doctrines for the employment of both RAF and Royal Navy tactical nuclear weapons were specifically set out within the NATO context. Briefly explained, the RAF's tactical nuclear weapons were integrated into the joint air and land battle plans of NATO and those of the Royal Navy into the anti-submarine warfare (ASW) defence of the NATO naval region around the United Kingdom. The British Army had access only to American nuclear weapons and these remained, theoretically at least, under absolute US control almost until the moment of firing. Thus the British Army only became a nuclear force within the NATO context. An additional point was that the British command, control and communications system which supported the use of tactical nuclear weapons was integrated into the NATO structure and, perhaps more importantly, was predominantly limited geographically to the NATO region.

The British were committed to the idea that in a situation of crisis or conflict authority over British nuclear weapons (after release under normal circumstances by British political authority) would be assumed either by US or NATO military commanders. However, the dismissal of any notion of national control over British tactical nuclear weapons ignored at least three important factors:

1 the integrity of British command within the NATO command structure;
2 the capability of the United Kingdom to order the use of nuclear weapons in Europe against the wishes of either the US or NATO; and
3 the capability of the UK to use tactical nuclear weapons outside the NATO theatre.

These points will form the basis for an examination of the tactical nuclear weapons command, control and communications systems of each of the three British armed services. Particular consideration will be given to the problems of maintaining centralised political control over these weapons and their potential impact on NATO's doctrine of controlled and limited nuclear war.

Even a brief explanation of the command and control of RAF tactical nuclear weapons cannot avoid dealing with the command structures in place to meet both national and NATO plans. The problems of enabling a peacetime national force to integrate quickly into NATO were dealt with by establishing two parallel command structures within which commanders wore both national and NATO hats.

The two elements of the RAF which were concerned with tactical nuclear strikes – Strike Command and RAF Germany – were commanded by British officers who simultaneously held NATO commands which they could assume in a period of crisis or during conflict.The Air Officer Commanding (AOC) Strike Command held the NATO position of Commander-in-Chief UK Air (CINCUKAIR) which was directly subordinate to NATO's SACEUR. Within Strike Command, No 18 Group – the maritime arm of the RAF – was commanded by an officer who held two NATO positions as Commander of the Air Channel (COMMAIRCHAN) and Commander of the Air East Atlantic (COMMAIREASTLANT), both of which were subordinate to NATO's Supreme Allied Commander Atlantic (SACLANT).[70] Finally, RAF Germany formed part of NATO's Second Allied Tactical Air Force (2ATAF) with Belgian, Dutch, West German and American aircraft. Despite the multinational nature of this force 2ATAF was always commanded by a British officer holding a NATO command subordinate to the Allied Air Force Central Europe (AAFCE) and ultimately to SACEUR.[71]

Thus although the tactical nuclear weapons of the RAF were committed to NATO, and although NATO would have assumed authority for their use in wartime, British officers (wearing NATO hats) remained in command of the weapons at all times. During crisis or wartime the only significant change for those officers would have been the 'chop' to direct subordination to NATO commanders rather than to national authority. Nonetheless a command structure which maintained British command over British nuclear weapons within the NATO framework must at least have allowed the British some autonomy in their nuclear actions. Although it may be hard to envisage any circumstances of independent British action, the question of to whom a British commander remained loyal if directed to one action by NATO and another by British political authority illuminates the real degree of RAF subordination within NATO.

British tactical nuclear aircraft (Harrier GR-5, Jaguar and Tornado) from Strike Command No 1 Group and RAF Germany which were committed to NATO's central front have been described as a 'contribution to SACEUR's theatre nuclear strike force'.[72] Within this strike force the aircraft had two main roles:

1 to disrupt Warsaw Pact forces until NATO's land forces were in position, and once this has been achieved
2 to disrupt the enemy army's battle plans.[73]

The RAF's only nuclear weapon (the WE177) had a yield of approximately 20 kt. This scale of destruction limited the potential target set against which the nuclear weapons could be directed being too large for attacks against targets in close proximity to friendly forces. The RAF's nuclear weapons were thus more likely to be targetted against large relatively static targets such as airfields, railway junctions or troop and tank concentrations deep within enemy territory.

This role and the fierce electronic hostility over the battlefield area had a significant impact on command and control of the RAF's tactical nuclear weapons. The aircraft which planned to execute nuclear strikes intended to do so, as far as possible, observing radio silence. To quote one RAF officer: 'If a sortie has been properly briefed and planned and is flown successfully there shouldn't be a need to say a word on the radio from start to finish'.[74] There was obvious sense in this approach since electronic jamming and other counter-measures could severely curtail any attempts at communications. In planning therefore the RAF could not anticipate assured communications, nor did they:

> . . . the RAF expects that in any major European conflict the degree of disruption and attrition to command and control systems will be so great that . . . squadrons . . . will be out of contact with higher headquarters for extended periods of time. Consequently, RAF procedures and training are based on the premise that central control will often be missing.[75]

Because of this training centred on nuclear strikes against single targets flown by lone aircraft which could not rely on assured communications with higher authority.[76] RAF pilots in the UK were trained to adopt the hi-lo-hi attack mode, flying high towards the intended target then hugging the ground to avoid radar detection and anti-aircraft missiles before delivering a nuclear strike and returning to high altitude for the return flight.[77] Nuclear strike aircraft of RAF Germany could not avoid the hostility of Europe's central region and therefore planned to adopt a terrain-hugging attack mode from the moment of takeoff. Aircraft leaving the UK for nuclear strikes trained to avoid the central European battlefield by flying northwards over Scandinavia before turning south to strike targets in Eastern Europe. During these sorties aircraft could perhaps have been recalled or redirected but this was not without problems for reasons already outlined.

Nimrods and Buccaneers, the nuclear capable aircraft of the RAF's maritime arm, No 18 Group, were tasked with anti-shipping and ASW attack in defence of the NATO naval region around the UK. During a crisis or conflict MR2 ASW Nimrods from RAF St Mawgan and RAF Kinloss were intended to assume a nuclear ASW role using nuclear depth bombs.

The RAF chose to rely on the Americans for the provision of nuclear depth bombs and in a conflict would have used some of the nuclear weapons stored at RAF bases at Machrihanish and St Mawgan. Under bilateral arrangements the RAF was not able to use these depth bombs without US authority. In all probability the UK would have acted jointly with US ASW aircraft operating in the same region. It seems probable that in a conflict armed US nuclear depth bombs would have been loaded aboard British Nimrods giving the British effective operational command of the weapons once release authority was given by the American President.

A second function of the Nimrods was to pass intelligence information to Buccaneer squadrons for anti-shipping and littoral land target nuclear strikes. Buccaneers used the RAF WE177 nuclear weapon which was not subject to US veto.

The RAF's efforts to integrate the command and control of both Strike Command and RAF Germany revolved in the 1980s around the Uniter communications network and the RAF command and control information system (UKAIRCCIS), a logistic and management system.[78] When completed these two systems were intended to provide communications and datalinks between all principal RAF stations in the UK and Germany based on commercial cable, troposcatter and satellite communication links between the two countries.

Of particular relevance to tactical nuclear operations were the RAF's mobile satellite links for use with the Skynet IV satellites. These provided direct communications between British command headquarters in the UK and RAF operational units within NATO, and were specifically intended for secure communications '. . . such as requests for nuclear release'.[79] By providing direct and secure means by which operational units could communicate with high-level national political and military authority the system if required could effectively have circumvented NATO's command structure.

On this point it seems clear that the RAF was capable of independent nuclear use within NATO, however unlikely this was considered to be. Both the command structure and the available communications systems allowed and supported the direction of the RAF's tactical nuclear operations by national command authorities if the need to do so arose.

The primary tactical nuclear role of the Royal Navy concerned the securing of NATO's resupply sea lanes across the Atlantic and the protection of the UK and British naval bases.[80] In effect this meant that the principal role of the Royal Navy was ASW operations in the North Atlantic, particularly in the Greenland-Iceland-UK Gap (GIUK) region and the North and Norwegian Seas.

As with the RAF the Royal Navy met both national and NATO obligations by operating two parallel command structures. As noted above the Commander in Chief, Fleet (based at Northwood) also held two NATO commands. To support these commands the British C-in-C Fleet had two subordinate operational commanders, one the Flag Officer at Plymouth, the other the Flag Officer at Pitreavie Castle. Each of these officers in turn held dual NATO posts. The Flag Officer at Plymouth was both the Commander of Plymouth Channel (COMPLYMCHAN) subordinate to CINCHAN, and Commander of the Central Atlantic (COMCENTLANT) subordinate to CINCEASTLANT and SACLANT. The Flag Officer at Pitreavie Castle was both Commander of the North Channel (COMNORCHAN) subordinate to CINCHAN, and Commander of the North Atlantic (COMNORLANT) subordinate to CINCEASTLANT and SACLANT.[81]

The Royal Navy operated both free-fall nuclear bombs and nuclear depth bombs from a variety of vessels, usually thought to include the Invincible class aircraft carriers, County, Type 42 and Type 82 destroyers, and Leander and Type 22 frigates.[82] The means of delivery of these weapons included Sea Harrier aircraft for the nuclear free-fall bombs and Sea King, Lynx, Wasp and Wessex helicopters for nuclear depth bombs.[83]

The Royal Navy relied on British nuclear weapons and made no use of American nuclear depth bombs (in contrast to the RAF's maritime arm). The main reason advanced for this was that the carrying of US nuclear weapons aboard British ships would have necessitated the presence of US nuclear weapons custodial units on board British vessels with physical control over the weapons in order to comply with US consitutional law. This arrangement was said to be impractical.[84] A second reason may have been that the Royal Navy had no wish to submit to a US veto on its tactical nuclear weapons operations. This may have indicated a desire for a tactical nuclear capability for use outside the NATO theatre. As one American commentator suggested:

> . . . the fact that Britain has nationally developed its own nuclear depth bombs for its helicopters suggest that the US could not supply these weapons to Britain. This may be because helicopter delivered nuclear depth bombs were primarily seen for non-NATO roles.[85]

Royal Navy command and control programmes such as the OPCON command and control systems and the ICS3 integrated communications system deployed in the 1980s supported high level and centralised control of nuclear weapons operations within the NATO theatre.[86] Once again the means were available to direct tactical nuclear forces along national lines should the need have arisen.

Unlike the other British armed services the Army did not have national nuclear weapons. Within the NATO alliance the Army did, however, have a nuclear role. The British Army on the Rhine (BAOR) fielded three nuclear capable delivery systems: the 105 mm and 203 mm artillery guns and the Lance missile system. Under bilateral agreements with the USA warheads for these systems were provided by the Americans during periods of crisis or conflict. Before exploring the nature of this arrangement more fully it is necessary to outline briefly the position of BAOR within NATO's command structure and to examine the national responsiveness of the force.

The BAOR defended a forward area in the north of West Germany as one of the four national armies in NATO's North Army Group (NORTHAG). NORTHAG in turn was one of two army groups [the other being the Central Army Group (CENTAG)] which together made up the land army element of the Allied Forces in Central Europe (AFCENT). AFCENT in turn was one of four major subordinate commands which were directly responsible to SACEUR.[87] Despite this command structure and the seemingly deep integration of BAOR into NATO the British Army retained an almost completely national identity because of the way in which national corps were organised along NATO's central front (see Figure 2.3 in Chapter Two). Because the NATO armies were organised on national lines with each army defending a specific region (or regions) operations and command tended to be national in nature. While command and control structures existed to enable NATO to coordinate BAOR operations there were also national links via the commercial, troposcatter (STARRNET),[88] and SATCOM carriers of the DCN to allow direct communications between BAOR and British authority.

Against this background an examination of the arrangements made for the provision of American tactical nuclear warheads for the British Army provides important insights into the processes and procedures for nuclear-sharing under NATO's bilateral nuclear arrangements. Under these plans the British Army provided nuclear capable delivery systems and the United States provided the nuclear warheads for those systems. In peacetime these warheads and delivery vehicles were kept remote from one another, as far apart, in some cases, as several hundred miles. In a period of crisis British delivery systems were intended to be dispersed into the field to be joined there by American custodial units with the warheads. This would obviously have created logistic problems which would not have been helped by the scale of mobilisation throughout the NATO theatre and the lack of fully interoperable communications.

The exact procedure for the transfer of American nuclear weapons to the British Army and the exact point at which American custodial units would have relinquished operational control of the nuclear weapons has not been disclosed. However, the information available on the arrangements for the Lance missile system does offer some insight.

The initial step in the operation of the Lance missile was that a high-security area would have been created in the field in which missiles and warheads could be mated. This area was intended to be protected by barbed wire (if there was time) and guarded by troops. Lorries would have brought the missiles and warheads separately (the warheads in US custody) to the secure area. The Lance missile system used two vehicles, the first, a tracked loader/transporter, was used to carry the missiles to the firing sites and also to load each missile onto the second launcher vehicle. Each missile would have been taken from the lorry and placed on the loader/transporter vehicle. Meanwhile a warhead would have been unloaded and the arming procedure carried out prior to the warhead being fitted to each missile. Once the warhead and missile were mated on the loader/transporter vehicle (this involved only four swing bolts) this vehicle and the self-propelled launcher vehicle were ready for front-line deployment together. At a suitable site in the field the armed missile would have been transfered to the launcher vehicle and coupled to the firing computer. Aiming and firing took about 20 minutes.[89]

This information does seem to indicate that once the missile was mated to the (armed) warhead the Lance was fully armed and therefore *de facto* released. It also appears that US custodial units were not intended to remain with the warheads once they have been mated with the missiles. In effect, therefore, the British Army would have had full operational control of the missiles at this point. If a similar de facto nuclear release took place with the supply of US nuclear artillery shells for British guns then the British Army could have had full operational control over its nuclear systems. Since dispersal and mating would have been likely to take place prior to hostilities (because of the risks of disruption) the British Army may have had full operational control of US nuclear forces prior to hostilities.

During the early stages of conflict the NATO command structure would have played a less significant role than the strictly national command structure. The problems of communications interoperability would have probably lead the British Army to defend its sector of the central front along national lines as vertical and horizontal communications with British commanders would be easier than with NATO allies. Problems of cross-national communications did raise doubts, within a NATO conflict scenario, about how SACEUR intended to retain centralised control over the use of nuclear weapons.

Even at its peak the British tactical nuclear arsenal comprised perhaps only 120 or so weapons in NATO's arsenal of many thousands. The small size of the British tactical nuclear arsenal consequently raised the question of what purpose it was meant to serve.

The primary rationale offered was that the weapons served as a contribution to NATO. The purpose and value of this contribution was not clear. Nor was it clear why the British Army considered it acceptable to be fully dependent on the Americans for nuclear munitions, while the RAF only allowed itself to be partially dependent and the Royal Navy did not allow any dependence at all. In all probability the answer to these questions lay outside either strictly military or NATO considerations.

The British Army, according to one source, '. . . adopted an atomic strategy even before nuclear warheads were available for its implementation'[90] and has relied upon American warheads since the beginning. The Heath government reportedly considered arming the Lance with British warheads but rejected the idea on economic grounds.[91] The RAF's tactical nuclear weapon – the WE177 – was originally deployed in the early 1960s around the time that the British government decided to hand the startegic deterrent over to the Royal Navy (Polaris). It may be that the RAF acquired the smaller weapon in an attempt to justify new nuclear roles. The Royal Navy seems to have eschewed any reliance upon American nuclear warheads in order to maintain operational independence. This suggests the desire to avoid an American veto and to retain the capability to use nuclear weapons outside the NATO theatre.

At least two suggestions have been made for independent tactical nuclear use by the British, the first to initiate nuclear use within Europe[92] and the second to thwart a cross-channel invasion.[93] The first of these seems to have lacked credibility. According to one source, such an act would only be concievable '. . . if the NATO alliance was already crumbling',[94] but if the alliance was crumbling it is difficult to see what purpose independent nuclear use by the British would have served. The second suggestion rested on the assumption that NATO's entire nuclear posture failed to prevent Western Europe being overrun and that the United Kingdom had not become embroiled in nuclear warfighting during the defeat of NATO.[95] This is hardly credible either.

Britain's tactical nuclear weapons would, however, have had a role in a NATO deterrent posture that was designed to threaten that any war in Europe could go nuclear, since Britain's nuclear forces could be guaranteed to complicate any notions of centralised command and control, thereby adding elements of uncertainty to any conflict. If this interpretation is correct then the posture was a dangerous one since it was based on the premise that any conflict could escalate to nuclear war and do so precisely in a way that would make it extremely difficult if not impossible to control.

CONCLUSIONS

An examination of political authority over the use of British nuclear weapons has indicated that the power of the Prime Minister was not absolute. Although the authority to use nuclear weapons rested solely in the hands of the Prime Minister, in practice (to meet the 'two-person' rule) the decision to use nuclear weapons would have been taken with at least one other individual. Moreover, the Prime Minister exerted no direct or physical control over the use of nuclear weapons beyond the considerable (and in peacetime paramount) authority of office. The ability to use nuclear weapons remained in the hands of the British military even if the authority to do so did not.

Furthermore, subject to the same caveats as the US President, the British Prime Minister was committed, if not bound, to consultation with his or her alliance partners prior to the use of nuclear weapons in NATO. The separate Anglo–US commitment to consult before the use of nuclear weapons by either party in any theatre further circumscribed British autonomy of action.

In the event of the death or incapacity of the British Prime Minister control of nuclear forces could have devolved to other political leaders (in a way that is less than clear) but ultimately would have devolved to the military forces themselves.

The assessment of the command and control of Britain's strategic nuclear forces pointed to a reliance on US early warning systems which did not in practice guarantee warning of attack for the United Kingdom. Moreover, the vulnerability of British command centres and the circumspect way in which the strategic nuclear forces were linked to political leaders raised the distinct possibility that the strategic forces could have been cut off (possibly without warning) from high level authority. In such circumstances nuclear retaliation could not have been guaranteed.

Since the British nuclear command and control system was vulnerable to even limited theatre nuclear strikes it was possible that such an attack could trigger strategic nuclear retaliation thereby ending any notions of escalation control in NATO.

The assessment of the command and control of British tactical nuclear weapons has shown that, as with strategic weapons, the military had *de facto* control of nuclear weapons if not the authority to use them. Moreover, there was a clear difference between the Royal Navy and Royal Air Force with respect to the autonomy and tone of nuclear operations. The British Army had a nuclear role only as one of NATO's bilateral nuclear forces. As such the analysis of British Army nuclear operations in this chapter provides a detailed consideration of how NATO arrangements for nuclear sharing could have operated in practice.

The absence of direct physical control of British nuclear weapons (strategic or tactical) by political leaders raised the possibility, in certain circumstances, that British nuclear weapons could have been instrumental in the loss of escalation control in NATO.

5 French Nuclear Command and Control

INTRODUCTION

Although outside the integrated military structure of NATO France is a member of the alliance. Moreover, as a nuclear power the role of French nuclear forces in the NATO theatre and the relationship between French and NATO nuclear command and control procedures and systems was central to an understanding of NATO nuclear command and control and the strategy of flexible response. This chapter stands as a case study but is intended both to inform and to be informed by the wider context of the discussion in this book.

This chapter is organised into five sections. The first presents a discussion of the nature of political authority over the use of French nuclear weapons. The second and third provide a description of strategic and tactical command and control systems respectively. The fourth discusses the extent to which these systems are able to support the French declaratory policy of proportional deterrence and the prestrategic function of tactical nuclear weapons. The fifth, drawing on some of the implications of the fourth, asks what the French nuclear command and control system implies about the French ability, and perhaps intention, to co-ordinate nuclear weapons operations with NATO.

POLITICAL AUTHORITY

Constitutionally, the question of who had ultimate control of French nuclear forces was unclear. Article 15 of the 1958 constitution of the Fifth Republic appeared to establish that it was the President of the Republic who had sole authority based on his position as Head of State, Chief of the Armed Forces, and President of the Defence Committee ('Le Président de la République est le chef des armées. Il préside les conseils et comités supérieurs de la défense nationale').[1] The actual text of the constitution was, though, ambiguous and internal contradictions seemed to muddy the waters still further. Article 20 noted that it was the government which had the 'use' of the army ('Le gouvernement . . . dispose de la force armée'), while Article 21 indicated that it was the Prime Minister who was responsible for national defence ('Le premier ministre . . . est responsable de la défense nationale')[2].

Decree 64-46 dated 14 January 1964 is more often cited as the basis for Presidential authority since it states:

> . . . the commander of the Strategic Air Forces is responsible for the execution of the operation of these [nuclear] forces on the order for engagement given by the President of the Republic.[3]

Analysts have been quick to point out, however, that at the time the decree was issued French nuclear forces consisted only of the nuclear gravity bombs carried by Mirage IV aircraft and that, by implication, the decree did not necessarily extend to those nuclear forces – S-3 missiles, submarine-launched ballistic missiles, Pluton tactical missiles, and tactical nuclear aircraft – which were subsequently developed.[4]

In practice the position was far clearer. Successive Presidents have declared that the authority to use nuclear weapons rested solely in their hands and pointed to the time-urgency of decisions to retaliate with nuclear weapons as effectively excluding consultation with Parliament or the executive.[5] The command and control arrangements (see below) which provided the incumbent President with the means to control both the strategic and tactical nuclear forces further underlined the primacy of Presidential office.

Presidential control of nuclear forces was exercised through the possession of an electronic code sequence 'key' which was prepared by the Elysée military staff. All French nuclear forces were protected against unintentional or unauthorised use by a series of safety and security measures, including procedural arrangements such as the two-person rule, mechanical switches and electronic locks (equivalent to PALs).[6] The President was able to 'unlock' the nuclear weapons by transmitting his unique code sequence. This code was passed down the various levels of military command in an unaltered form and physically unlocked the weapons *in situ*.[7] This meant that the Presidential code was also an enabling code.[8] Moreover, the code acted simultaneously to sanction the use of the weapons which were released, a function which was adapted to the general release of nuclear weapons envisaged by the French proportional deterrence strategy.

For security purposes the Presidential code was changed frequently. It has been reported that prior to President Mitterrand's period of office French Presidents were obliged to memorize this code, though this seems highly imprudent. When Mitterrand took office it was decided that he should be accompanied at all times by an officer from the Elysée military staff who would ensure that he remained in continuous contact with the nuclear forces.[9] It is understood to be the case that the President carried the code in the form of an electronic card 'key'. This code and targeting information could then have been rapidly transmitted to the nuclear forces with the assistance of the accompanying officer.

A further feature of Presidential control was thought to be the ability to deactivate (i.e. 'relock') certain nuclear weapons which had been unlocked by prior Presidential authority. This capability was reported for aircraft-delivered nuclear weapons and may have extended to other weapons.[10]

The obvious flaw in this arrangement of centralised presidential control was the same threat of decapitation that has been described for American (and British) nuclear forces. Arrangements must therefore have also existed in France for some form of command delegation and it is known that in the event of the President's death or incapacitation the Prime Minister was able to assume his office and direct nuclear operations.[11] However, even this provided little real redundancy since both individuals were likely to be in close proximity at a time of international tension or war. It was therefore likely, though there is no explicit evidence, that in the event of a seriously deteriorating international situation steps could have been taken to distribute the critical Presidential code to a number of senior nuclear commanders.[12] Because of the risks of unauthorized use of nuclear weapons further delegation to lower command levels seems improbable.

Another area of Presidential primacy in controlling nuclear weapons was in the targeting of nuclear forces. The President had the duty of selecting targets which reflected French deterrent policy[13] from a catalogue prepared by the nuclear commands. He was assisted in this selection by his senior military council ('conseil de défense').[14]

In directing the release and use of nuclear weapons the President had available a command centre which was been built in the basement of the Elysée Palace.[15] Code-named 'Studio Jupiter', the command centre hosted a range of data-processing, display, teleconferencing and other communications facilities.[16] The President was linked to France's National Military Command Centre at Taverny (Val-d'Oise) just outside Paris by a number of communications systems. These included three primary hardened systems code-named 'Ivory', 'Jupiter' and

'White' which were backed up by commercial telephone lines (Poste, Télégraph et Télécommunications [PTT]) and a radio link.[17] Similarly redundant communications also linked 'Studio Jupiter' to the Alternative National Military Command Centre at Mont Verdun (Lyon) and to the Ministry of Defence at Rue St Dominique in Paris.

Finally to confirm the sending of the Presidential code there was a special validation system linking 'Studio Jupiter' to Taverny and Mont Verdun. By means of a television link which established the presence of the President and telewriting – which was thought to transmit a real-time picture of the Presidential signature as it was being written (which could be compared with samples of the signature already held) – military personnel in the command centres were able to verify that the President had indeed issued the code for nuclear release.[18]

STRATEGIC NUCLEAR COMMAND AND CONTROL

Operational control of strategic nuclear forces was exercised from the Centre d'Opérations des Forces Aériennes Stratégiques (COFAS).[19] Buried 60 metres beneath Taverny on the outskirts of Paris, and constructed of concrete, the COFAS integrated three essential command and control functions. Firstly, it served as the senior military command centre for nuclear operations and could host the President and other key political leaders for the period of critical decision-making (in the event that the President chose to leave the Elysée Palace). Secondly, it was the headquarters of the Forces Aériennes Stratégiques, the French strategic airforce, which had operational control of the Mirage strategic bombers and the S-3 ballistic missiles. Moreover, although France's strategic ballistic missile submarines were controlled in peacetime from the Centre d'Opérations de la Force Océanique Stratégique at Houilles, in wartime control passed to COFAS.[20] Thirdly, Taverny also co-located the Centre d'Opérations de la Défense Aériennes (CODA) which controlled and integrated air defence and attack warning.

The facilities at Taverny were duplicated at the Alternative National Military Command centre at Mont Verdun near Lyon. This underground command post was capable of fully exercising the functions of Taverny in the event of the destruction or degradation of the latter.[21]

The first indication of an impending attack against France would probably have come from US intelligence sources. France herself had virtually no intelligence-gathering satellite capability and would certainly have had to rely on the US and NATO for warning of Soviet preparations for attack. France deployed a primarily civilian observation satellite system, SPOT (Satellite Pour l'Observation de la Terre), but this had limited military utility because of its relatively poor optical resolution (20–30 metres), and lack of night-time or cloud penetration capability.[22] Attempts begun in the early 1970s to develop an intelligence satellite capability, which led to the SAMRO (Satellite Militaire de Reconnaissance et d'Observation) project, were eventually abandoned in the early 1980s.[23] In 1986 the project was resurrected under the Helios satellites programme which had minority participation by Spain and Italy.[24] The satellites were reported to be intended to carry relatively high resolution optical observation equipment (resolution 2–3 metres) and to carry electronic listening devices and radar for all-weather and night observation.[25] Helios was, however, not deployed by 1990 and the programme is therefore not germane to the present discussion.[26]

At first glance France also appeared to have little or no ballistic missile warning capability to detect the launch of Soviet missiles or to track them in flight. At least one source notes that the French air defence network was linked to the US Ballistic Missile Early Warning System, though whether this was directly from NORAD, the US fusion centre for attack warning, or via NATO command centres, is unclear.[27] However, France had at least a rudimentary ballistic missile warning apparatus herself. The regular flight testing of missiles off the French coast and the robust Ariane space programme suggest that France had a well developed radar target acquisition and tracking capability. The family of radars developed for missile tracking purposes by Thomson-CSF, including the Ardour II, Artois, Bearn, and the new Atlas, were deployed in France at a number of sites and were certainly capable of providing some form of real-time attack.[28] No evidence has been found, however, that these radars were linked to any central attack warning system.

There was no such ambiguity about the French air defence network. The STRIDA (Systeme de Traitement et de Representation des Informations de Défense Aériennes) system began development in 1956 and was integrated with the NATO NADGE and British UKADGE air defence networks. The STRIDA was a network of ten main radar stations linked with Centaure and Aladdin radars as well as the civilian air traffic control network. Information from the entire system was fed into the CODA at Taverny and the back-up facility at Mont Verdun. In the 1980s France agreed the purchase of four E-3 AWACS aircraft from the US to supplement the STRIDA network.[29]

Nevertheless, throughout the period under discussion France was largely dependent upon allied warning systems. By 1990 the situation has changed little from that in 1980 when a leading French defence commentator noted:

> . . . if we did not have at our disposal information from American radar networks and from NATO forces with NADGE (high altitude) and with AWACS (low altitude), we would be practially blind.[30]

Routine military communications in France utilised a military network which was based on the national civilian PTT network. The network provided telephone and telegraph (message distribution) facilities for most land-based military users. The system was backed-up by the PTT network itself.[31] The primary communications network between Taverny, Mont Verdun and subordinate nuclear command centres was a secure and protected analogue network known as RA-70. It was on this network that the release code and orders were intended to be passed to the nuclear forces.[32]

In April 1988 the French Ministry of Defence took delivery of the RAMSES (Réseau Amont Maillé Stratégique et de Suivre) communications system.[33] RAMSES was a multiservice digital network which carried telephone, telegraph and data transmissions and was reportedly 'fully protected and hardened'.[34] The full operational capability of the system was not due to be exercised before 1996.[35] RAMSES was intended to transmit the nuclear release code and orders from the President via Taverny (or Mont Verdun) directly to the S-3 (later S-4) missile command centres at Rustrel and Reilhanette on the Plateau d'Albion, to the Mirage bomber bases,[36] and to the two main VLF transmitters at Kerlouan and Rosnay for onward transmission to the ballistic missile submarines (an ELF transmitter was also under development at Rosnay). These two VLF transmitters were backed up by other VLF and LF transmitters at Brest, Cherbourg, Croix d'Huis, La Régine, St Assise, and Toulon.[37] Further submarine communications redundancy was reportedly possible through the use of VLF antennae trailing from hot air balloons, though this would presumably only have been used in an emergency.[38]

The question of the vulnerability of vital VLF transmitters prompted the French Government in 1982 to order four airborne VLF command post facilities from the US. Adapted from the American TACAMO system the command posts were installed on French Transall C-160H aircraft.[39] The first aircraft in the programme known as ASTARTE (Avion Station Relais de Transmissions Exceptionelles) became operational in February 1988[40] and the programme was completed by the end of 1990.[41] The limited number of aircraft meant, however, that there was no peacetime effort to maintain an ASTARTE aircraft permanently airborne, although such a capability was intended to be exercised in periods of crisis. All four aircraft were based at Evreux and operated by the airforce transport command COTAM (Commandement du Transport Aérien Militaire), although operational control was exercised by the COFAS.

The final important element in strategic communications was the Syracuse satellite communications network. The Syracuse I network was based on a military package aboard the predominantly civilian Telecom 1 satellites. The first two Telecom satellites, 1A and 1B, were launched in August 1984 and May 1985 respectively.[42] From geostationary orbit both satellites provided identical coverage of the area from the Middle East to the eastern coasts of North and South America.[43] The Syracuse military ground network comprised 26 stations including three fixed sites at Brest, Chartres and Castelnaudary, nine air-portable land stations, three light mobile land stations, and 11 stations on board vessels of the French Navy.[44] Syracuse I provided real-time communications to users and was one of the principal systems used by the French Navy operating in the Gulf during the Iran-Iraq war to maintain links with fleet headquarters and Paris.[45]

In the late 1980s the Syracuse I network ran into difficulties. In late 1987 the Telecom 1A satellite suffered technical problems and the bulk of communications traffic had to be switched to the 1B satellite. On 15 January 1988 the French lost attitude control of the Telecom 1B satellite, rendering it effectively inoperable. Communications were switched back to the 1A satellite which, according to government officials, continued to provide adequate reliability.[46] Some of the burden was assumed by the Telecom 1C satellite launched in March 1988, and the Syracuse I programme was scheduled for replacement in the 1990s.[47]

The Syracuse II programme, based on two Telecom 2 satellites[48] promised greatly to improve the French satellite communications capability extending satellite communications to the air force – which to 1990 had no such capability – most notably to French AWACS and to the ASTARTE Transall aircraft.[49]

TACTICAL NUCLEAR COMMAND AND CONTROL

As with the strategic nuclear forces all tactical nuclear weapons were under the direct authority of the President and were controlled through the Chief of the General Staff. Under the Chief of the General Staff's direction, operational control of nuclear forces was exercised by the commanders of the three services.

The commander of the 1st Army, based at Strasbourg, controlled the Pluton tactical nuclear missile system. Pluton missiles were fired from a mobile tracked launcher. Command and control vehicles accompanied the launcher and carried both communications and data-processing equipment.[50] Target data was acquired either through observations by other military units or through the use of R20 drones which used infrared sensors and transmitted battlefield information in real-time to the Pluton launch computers.[51]

Plans were made in the 1980s for the Pluton missile system to be replaced by the longer-range Hades system. Hades was intended to be similarly mobile, although with launch and command and control functions integrated into a single vehicle so that launch crews need not leave the vehicle's cab to execute missile launches.[52] Hades was designed to use information from a pilotless reconnaissance drone, the CL289 developed specifically for the purpose, for target acquisition. Hades was also designed to utilise an advanced version of the existing RITA (Réseau Intégré de Transmissions Automatiques)[54] battlefield communications network to provide protected communications between command centres and the Hades launchers deployed in the field.[55] Hades did not however begin operational deployment until 1992 and is therefore not relevant to this study.[56]

Nuclear weapons of the French tactical air force were controlled from the FATAC (Force Aérienne Tactique) headquarters at Metz. The aircraft were deployed in dispersed hardened (above ground) shelters and maintained at armed readiness.[57] Dedicated communications were maintained with the aircraft enabling nuclear weapons on board to be released prior to or shortly after takeoff. Electronic hostility closer to the battlefield area risked interference with communications links and argued against delaying release for too long after takeoff. The RAMSES programme was extended to tactical nuclear airbases, thereby strengthening high level control of the aircraft.

Naval tactical nuclear weapons aboard the *Clemenceau* and *Foch* aircraft carriers[58] were directed from the Naval Mediterranean Command at Toulon. Both vessels deployed Syracuse satellite terminals to enhance communications with command centres and Paris. In the late 1980s the French Navy introduced a new high-frequency communications system linking surface vessels to shore facilities. According to one report: '. . . the new system will bring the French Navy a fail-safe means of communications with any point of the globe, attaining one of the highest protection factors of any HF signalling system in the world'.[59]

NUCLEAR COMMAND AND CONTROL AND FRENCH DETERRENCE

The credibility of French deterrence rested, inter alia, on the ability of the nuclear command and control system to realize the threats implicit in French declaratory policy. At the strategic level the doctrine of proportional deterrence demanded assured nuclear retaliation under any realistic circumstances at a level consistent with the stated aims of French deterrence. Moreover, it required the prompt and synchronised use of all surviving elements of French strategic forces in a single 'enlarged countercity' blow which made few demands on missile accuracy and rejected limited or counterforce options.[60]

At the tactical level, whilst rejecting the graduated use of nuclear weapons envisaged by NATO's strategy of Flexible Response, French forces had to be sufficiently flexible and robust to threaten the range of targets deemed necessary by French leaders to signal French resolve and to test the will of the Soviet Union to continue.[61]

The prestrategic function of tactical nuclear weapons was reflected in the employment concepts articulated for their use. Two types of targets were described for tactical nuclear weapons. The first, termed targets of opportunity, included front-echelon tank, troop, and artillery concentrations. the second, termed interdiction targets, included fixed infrastructure targets such as airfields and rear troop and tank concentrations massing for reinforcement.[62] The French envisaged a single closely synchronised attack against a range of opportunity and interdiction targets designed to arrest the progress of advancing conventional forces.[63] Crucially, the employment of tactical nuclear weapons was stated to be more '... orientated to the political management of a crisis than to military effectiveness',[64] placing a premium on maintaining political control of the nuclear forces.

Whether these requirements could have been met by the French nuclear command and control system is a difficult question to answer. To deal firstly with the question of whether the French command and control system could have assured nuclear retaliation it is necessary to look more closely at the strengths and weaknesses of the system. The potential vulnerability of French strategic nuclear command and control to a variety of threats had long been recognised and provided the impetus behind calls for improvements in redundancy and survivability.[65]

An obvious vulnerability, for reasons already outlined, was the centralisation of command authority in the hands of the President. Such a vulnerability may actually have been greater in theory than in practice since the President could devolve control as far down the command chain as he deemed necessary in order to avoid the threat of decapitation. Moreover, unlike the NATO strategy of Flexible Response which was premised on political control during nuclear exchanges, the French strategy of a single massive retaliatory launch of nuclear weapons was not compromised by the devolution of command authority, providing sufficient cohesiveness remained to coordinate elements of the nuclear forces.

Direct threats to the strategic command and control system itself were perhaps a more significant vulnerability. An electromagnetic pulse (EMP) generated by a nuclear detonation high above France, or even outside French territory, was taken particularly seriously[66] because it could have destroyed or degraded data-processing, communications and sensor equipment thereby inhibiting nuclear operations without the need to directly attack or invade France.

The principal nuclear command centres, the S-3 missile silos and launch control centres on the Plateau d'Albion and the Mirage IV bombers dispersed at numerous bases around France were also vulnerable to direct nuclear attack. Although the S-3 missiles could reportedly have been launched from a high level of alert in around 3.5 minutes[67] and the Mirage bombers ordered to take off and fly towards aerial holding points before receiving the release and use authorisation code,[68] neither was assured of escaping destruction in a nuclear attack. Flight times of Soviet missiles in Europe were so short as to provide minimal, if any, warning time to French forces and it must be doubted that sufficient time was available for command centres to communicate release and use authorization codes to the nuclear forces before the command centres were themselves destroyed. Even if the Mirage IV bombers could have taken off before missiles struck they would have been carrying nuclear weapons 'locked' and inoperable because codes were not received from the command authority.

It was possible to circumvent this vulnerability by the prior release of nuclear weapons but this would have implied a separation of the release and use authorization function which the Presidential code was understood to confer.

The French reportedly eschewed any planning for launch-under-attack or launch-on-warning,[69] suggesting that, short of first use by the French, the land-based elements of the strategic forces were not expected to survive to carry out their function.

Even a relatively small number of Soviet nuclear weapons could, with no or practically no warning, have destroyed the command centres at the Elysée Palace, Taverny and Mont Verdun and those at the principal subordinate sites, effectively severing those forces from high-level political and military control. The devolution of command authority could not have fully addressed this vulnerability given the expected destruction of the land-based forces. Predelegation of release and use authorization would have had the greatest utility when applied to the third and most survivable element of French strategic nuclear forces, the ballistic missile submarines.

It was upon the submarines that the ultimate burden of French nuclear retaliation rested and it was in submarine command and control that the heaviest investment was made. The ELF, VLF and LF transmitters previously described were vulnerable targets which could not be significantly hardened. The ASTARTE airborne VLF command post programme thus emerged as the principal element in assuring submarine communications.

The Transall C-160H aircraft which carried ASTARTE VLF equipment were reportedly able to remain airborne for 10 hours without refuelling and considerably longer with airborne refuelling.[70] In a crisis one of the four aircraft in the programme would have taken off from Evreux and upon reaching the necessary height would have deployed an antenna several kilometres in length whilst it circled above France. By rota, one aircraft taking off and assuming operation before the other left its station and returned to Evreux, the programme intended to maintain a permanent airborne VLF presence during a period of crisis. Aside from the high survivability of such an aircraft the ASTARTE command post also had the abliity to store the release and use authorisation code if so delegated by the President. This enabled the aircraft to transmit the release and use code to the ballistic missile submarines even if command centres on the ground were destroyed. Since most sources indicate that at least three submarines were usually operational at any one time[71] and one or perhaps two others could have been made ready to launch nuclear weapons from port or refit, at least 48 and possibly as many as 240[72] warheads could have been available.

Some question marks existed about whether in practice the ASTARTE programme could have assured nuclear retaliation. The first was the risk that sabotage of the aircraft – perhaps in the course of a conventional war – may have deprived France of their use. The second was that an EMP detonation may have inhibited their ability to function (even though the aircraft were EMP hardened this did not guarantee immunity from EMP effects). The third was that in a prolonged period of crisis a fleet of just four aircraft would not have been sufficient to guarantee a permanent airborne presence. The fourth was that the quality of VLF transmission may have been inadequate, particularly in a hostile environment, to properly communicate the release and use authorization code. The latter was particularly important because French submarine-launched ballistic missiles were equipped with electronic 'locks' (PAL-type systems) and an incomplete or degraded signal may have meant crews could not unlock the weapons. A report that the two powerful land-based transmitters at Kerlouan and Rosnay needed to work in unison to assure smooth message transmission to the submarines[73] did not auger well for the clarity of a transmission from a single and much less powerful ASTARTE aircraft.

Nevertheless, in all but an unexpected bolt-from-the-blue attack – in which the centralisation of command authority emerged as an Achilles' heel because there was no time to devolve the Presidential code – the diversity of French nuclear forces and of strategic command and control was probably enough to underwrite some level of nuclear retaliation.

At the tactical level the ability of French command and control to support the employment concept envisaged by the prestrategic function of the weapons was greatly enhanced by the fact that most of the weapons themselves were intended to remain on French territory away from the forward battle area. The fact that German territory – whether French conventional forces are fighting there or not – served as a buffer against the conventional overrun of French territory provided France with the necessary time and space to direct tactical nuclear weapons to maximum effect. The dispersal of French tactical nuclear aircraft and the mobility of Pluton argued against their successful targeting by conventional forces. Thus French tactical nuclear weapons, like those in rear areas elswhere in NATO, would probably have remained inviolate for a period.

It is important to note that command and control difficulties would have become apparent if France had attempted to take nuclear weapons forward into the battle area. The deployment, for example, of Pluton missiles on German territory would firstly have incurred command and control penalties because of the difficulties of establishing and maintaining communications with Paris, and secondly would have exposed the missiles more directly to the hostility of the battle area (including electronic warfare) and the risk of overrun. In such a situation circumstances might not have permitted the synchronisation of the Pluton missiles with other tactical nuclear forces as envisaged by the prestrategic function. One can therefore ask whether there could have been a justification for taking Pluton forward into West Germany which would outweigh the potential inhibition of the prestrategic function of the weapons.

A further question mark existed over the selectivity of tactical nuclear forces and whether, with the stated aim of crisis management (and the need for assured political control which that implies) in mind, the forces could have been made a sufficiently exact instrument to meet deterrent needs. By specifying a release package of predetermined size which had to be used within a given area and a given time (to attempt to synchronise the prestrategic function) the President exerted a degree of control over nuclear weapons operations. Once released, however, control of those forces directed against targets of opportunity would inevitably have passed down the command chain because of the need for real-time responsiveness, and the ultimate decision to use the weapons would have rested with field commanders in a fluid battle area. The scale of the employment of nuclear weapons envisaged by the prestrategic blow may have presented enormous difficulties of coordination for French forces, perhaps leading to the time-staggered and incomplete employment of the weapons.

If France had been pushed to the point of using tactical nuclear weapons by the prospect of imminent invasion it is at least possible that in defeat elsewhere in Europe NATO would by then itself have had recourse to nuclear weapons. Quite how France would have coordinated her own forces in such a context to deliver a prestrategic blow which could be perceived as such is difficult to determine. It seems rather more likely that the use of large numbers of tactical nuclear weapons by France would have further plunged NATO into an escalatory spiral in which notions of political control would be lost. One alternative, that of restraint and coordination with the NATO allies, suggests a degree of nuclear cooperation which hitherto France has openly rejected. Another, the attempt to sanctuarize France by witholding nuclear forces whilst seeking a deal to end hostilities implied either the abandonment or the politically traumatizing withdrawal of French conventional forces from Western Germany.

Although France rejected any automatic integration with the NATO military command structure and insisted that the decision to use French nuclear weapons was a decision for France and France alone, there was nothing in French declaratory policy which prevented political and military coordination of nuclear operations with the NATO allies. The decision to use any nuclear weapons in Europe would have had such monumental consequences for all the NATO allies that it is almost certain that France and NATO would have coordinated their actions to some degree.

FRENCH NUCLEAR COMMAND AND CONTROL AND NATO INTEGRATION

It serves to clarify the relationship between France and the NATO alliance to first note that, as described in Chapter One, even within NATO's integrated military structure itself the allies were committed by no more than their own willingness to participate in the alliance.[74] Whilst political and military bonds were strong, in the darkest of circumstances when differing perceptions of national responsibility in the face of war emerged most clearly, the NATO alliance itself could have fragmented.[75] French intentions to act within or outside the alliance and the nature of any cooperation would have depended on the circumstances prevailing at the time, a rationale which accounted for the French rejection of any automatic commitment to NATO's military structure, and one which was no more than an overt statement of the realities facing the other NATO allies.

No one can say for certain how France (or any other nation) would have acted in a conflict between the Warsaw Pact and NATO in Europe. For present purposes one can, however, by considering French command and control arrangements, attempt to consider to what extent France was capable of coordinating her nuclear forces with NATO.

At the political level France retained a permanent representative and national delegation on the North Atlantic Council (NAC), NATO's most senior political decision-making forum.[76] The NAC, according to NATO documents, would in a time of crisis have become the principal forum for consultation between allies on the management of the crisis and the politico-military issues which arose from the situation. France would thus have had an equal voice at the highest political level of the alliance.[77]

To facilitate the exchange of information in times of crisis and war France joined the NATO-wide Communications System in 1979 (10 years after it became operational).[78] France has, however, taken no significant part in the more ambitious NATO Integrated Communications System. To supplement the usual means of international telephone communications between senior political leaders, France established a direct and protected 'hot-line' to the German capitol for the express purpose of consulting more closely in times of crisis.[79] Furthermore the French President spoke at a Franco-German summit in February 1986 of the possibility of consulting with West Germany on the use of tactical nuclear weapons:

> Within the limits imposed by the extreme rapidity of such decisions the President of the French Republic is prepared to consult the Chancellor of the Federal German Republic on the possible use of French prestrategic weapons on German territory. He reminds us that in this matter, the decision cannot be shared.[80]

At the military level cooperation between France and NATO appears to have been closer than past evidence has suggested. According to an article which forced a reassessment of Franco-NATO, and particularly Franco-American, military cooperation, the author noted that: 'The French armed forces were never happy with the break with NATO that de Gaulle imposed on them in 1966, nor with the declaratory nuclear doctrine that flowed from the break.'[81] Accordingly, steps were taken to ensure that France could continue to fight alongside her NATO allies:

> Successive French chiefs of staff and the American generals serving in the combined roles of SACEUR (Supreme Allied Commander, Europe – the chief NATO military officer) and CINCEUR (Commander in Chief, Europe – the commander of US forces)

> always kept open lines of communication and worked quietly to assure that even though France was not a member of the alliance's integrated military structure its forces could effectively participate in the defense of the West.[82]

The effectiveness of this relationship might have been behind the confident assertion of General Bernard Rogers during his term of office as SACEUR that he had '. . . always been convinced that, if NATO were attacked by the Warsaw Pact, the French would join the allies in the defence of Western Europe'.[83] Such a level of cooperation was no more than consistent with President Mitterrand's own stated view that '. . . in a conflict where there was immediate danger, the duty of France would be to come to the aid of our allies'.[84] What was rather less certain was how far this cooperation had been extended at the nuclear level.

Prior to their withdrawal from the integrated military structure of NATO French military officers regularly attended meetings of the American Joint Strategic Target Planning Staff.[85] At these meetings the targets for strategic nuclear weapons were discussed and efforts were made to coordinate targeting to avoid wasteful cross targeting by the US, the United Kingdom and France. It is likely that, as with the United Kingdom, France at the time developed two strategic targeting plans. The first a national plan designed to serve purely national purposes and the second a plan coordinated with NATO. When France withdrew from NATO the second NATO plan may have been abandoned or some reduced level of coordination may have continued.

Either way, some level of target coordination was subsequently re-established. The move away from the Gaullist line initiated by President Giscard d'Estaing in the mid-1970s led to improved relations with NATO and permitted the discussion of nuclear targeting by the then SACEUR Alexander Haig and the French Chief of Staff Guy Mery.[86] Despite the subsequent return to Gaullist verities in declaratory policy during the Mitterrand presidency the relationship moved, if anything, closer still. By 1985, following the highly unusual sale of Cray computers to France by the US, there was strong speculation that targeting cooperation was again taking place.[87] Evidence suggested that this coordination was taking place at the tactical rather than the strategic level and that the discussions were being carried out between France and NATO's SACEUR rather than with the JSTPS. According to this evidence NATO was concerned that the nature and scale of the French prestrategic blow could have escalated a conflict in Europe at a time when NATO was seeking to control the situation. France was apparently willing to coordinate a tactical plan with NATO (by exchanging target lists) in order to facilitate the desired restraint but did not compromised the prestrategic function of her own forces.[88]

In the broadest sense cooperation between France and NATO in military command and control infrastructure was considerable. In 1988 the French defence minister spoke of the growing interoperability between French and NATO systems and of an increased French willingness to cooperate in these areas.[89] Aside from French reliance on NATO and the US for missile attack warning and air defence, which has already been discussed, France cooperated with NATO in almost every military environment in Europe. To cite but a few examples, France participated in the BICES programme to integrate NATO battlefield intelligence,[90] the ACCS project to link air command and control,[91] and the ACE HIGH communications network.[92] She was also involved in plans to integrate the RITA battlefield communications system with those of the NATO allies and in particular the German AUTOKO system,[93] and to integrate the air, land and sea information distribution system SINTAC (Système Intégré de Télécommunications Tactiques) with the US JTIDS system developed throughout NATO.[94] Finally, she utilized (through the Syracuse programme) satellites of the NATO allies,[95] and participated in the US GPS/Navstar satellite navigation project (developing its capabilities for air, land and sea vehicles).[96]

The level of French coordination with NATO allies in these areas was comparable with that of the other allies. Moreover, French willingness to place at least the French 1st Army and the FAR in Germany under the 'operational control' of West Germany[97] suggested a level of command integration which might also, under certain circumstances, be consistent with that of the NATO allies.

At the nuclear level, however, such close command system integration did not take place (though some of the systems discussed above had utility for nuclear command and control) nor were any arrangements made to subordinate French nuclear forces to NATO command. The means for high-level political coordination of the use of nuclear weapons were in place, as were 'secure and direct' crisis communications between the French Chief of Staff and NATO's SACEUR[98] to facilitate military coordination of nuclear strikes. The apparent tension between NATO's strategy and the French prestrategic function could have evaporated if the decision to use nuclear weapons had ever been taken. Both NATO and the French would have been intent, if declaratory policy provides a guide, on employing nuclear weapons for the same purpose of blunting a Soviet conventional attack and signalling resolve. French coordination with NATO in such a strike was entirely consistent with the prestrategic function. If the Soviet Union retaliated with nuclear weapons NATO's own studies showed that neither side could thereafter accrue definite advantage from nuclear escalation[99] and there were doubts about the control of nuclear escalation after that point in any case.

According to French strategy in the event the first use of nuclear weapons failed to meet its political objective and invasion or nuclear strikes were launched against France, French strategic nuclear forces would have been used, according to the doctrine of proportionate deterrence, linking the fate of the US to that of Europe by the strongest of bonds. Whether this preceded, was coordinated with, or followed the use of strategic nuclear weapons by the US would probably have been irrelevant for France, which in any circumstance would have faced destruction.

CONCLUSIONS

In France the President exercised a degree of personal control over French nuclear forces considerably beyond that of either the US President or the British Prime Minister. His autonomy of decision-making was virtually unlimited and France was party to no formal commitments to consult with any other parties prior to the use of nuclear weapons (the possible exception being Mitterrand's indication of a willingness to consult with Western Germany prior to using tactical nuclear weapons, but this was not be considered a binding commitment), though of course he may have chosen to do so.

Presidential authority was exercised through authorisation and release codes which passed unchanged down the command chains to the nuclear forces. Because of the weaknesses of this tight centralisation of control the President was able to delegate his codes as required and in the event of his death or incapacitation could have been succeeded by the Prime Minister. In practice, however, authority would have been likely to pass to senior military leaders.

France deployed a scaled-down nuclear triad: ballistic missile submarines, land-based missiles and long-range bomber aircraft. The command and control apparatus in place to support these weapons was dedicated and, for the most part, robust. Nevertheless, both French nuclear forces and the nuclear command and control infrastructure were vulnerable to degradation or destruction. Ultimately French retaliation – and hence the credibility of her deterrent posture – rested on her submarine force and the ASTARTE airborne command posts deployed to provide assured communications with the submarines. Under all but the most implausible circumstances some form of French nuclear retaliation could probably have been achieved.

The assessment of French theatre nuclear command and control systems indicated a high degree of co-ordination with NATO, including the joint development and operation of certain command and control systems. France was not, however, committed to joint nuclear operations though the means existed to enable this. Plausible scenarios could be foreseen in which France would have had the motive to co-ordinate nuclear operations with NATO particularly in the early stages of conflict. Yet as the conflict progressed the disparity between French and NATO strategy and the possibility of French resort to strategic nuclear weapons would clearly have threatened any efforts by NATO to control and limit a theatre nuclear war. French nuclear forces and their command and control vulnerabilities both complicated nuclear operations in the NATO theatre and heightened the risks of nuclear escalation.

6 Political Decision-Making in Crisis and War

INTRODUCTION

The political control of NATO's nuclear weapons was central to the strategy of Flexible Response. Initiating the use of nuclear weapons and directing those weapons to meaningful political purposes required express political authority for, and close political control of, nuclear operations. Consequently it is important to explore both the role political decision-makers had in the command and control of nuclear forces and the effects the command and control system exerted on the political decision-making process.

For the political leaders of NATO the two questions having most direct bearing on the command and control of nuclear forces were firstly whether to commit national military forces to operational NATO command under the Chop procedures and secondly whether to agree the use of nuclear weapons if circumstances were thought to require their use.

In terms of the command and control of nuclear forces in NATO the speed and unity with which these decisions could be made by alliance members would have been crucial questions.

This chapter attempts to address these questions by looking at the nature of operational decision-making in NATO during crisis and war. It has been organised into six sections. The first considers the nature of national and coalition decision-making at the political level and primarily asks whether operational decisions could have been taken collectively by the alliance. It, however, also seeks to understand how national and collective processes of political control of nuclear weapons may have inter-related and to assess the effect of this for timely, accurate and coherent decision-making.

The second focuses on the decision to commit national forces to operational NATO control. It attempts to explore the tensions and dilemmas inherent in this crucial early decision.

The third looks at how, and on what basis, the decision to use nuclear weapons could have been taken. It explores the different aspects of the decision by discussing the political goals of alliance member nations and the nuclear options available to meet those goals.

The fourth considers the political aspects of the second use of nuclear weapons. It is premised on the failure (or partial success) of initial use to achieve NATO's political aims and seeks to understand how political decision-makers would have addressed the question of the aims and options of authorising a second nuclear strike.

The fifth section considers the political-military interface and looks at the dilemmas and problems created by operational political-military tensions, the problems of matching military actions to political goals and the influence military operations may have exerted on the political decision-making process.

The sixth and final section deals with the military issues raised by analysis of command performance of NATO nuclear weapons infrastructure in crisis and war and shows how these complicate further the political and politico-military issues discussed in this chapter.

NATIONAL AND COALITION DECISION-MAKING

Theories of alliance formation, behaviour and disolution provide a useful framework within which to consider the nature of national and coalition decision-making in NATO. According to Morgenthau:

> A typical alliance attempts to transform a small fraction of the total interests of the contracting parties into common policies and measures. Some of these interests are irrelevant to the purposes of the alliance, others support them, others diverge from them, and still others are incompatible with them. Thus a typical alliance is imbedded in a dynamic field of diverse interests and purposes. Whether and how long it will be operative depends upon the strength of the interests underlying it as over against the strength of the other interests of the nation concerned.[1]

The cohesiveness of an alliance, which is central to the present discussion, has been usefully summarised by Liska as resting on 'the relationship between internal and external pressures, bearing on the ratio of gains to liabilities for individual allies'.[2]

The relevant questions raised by these theoretical summaries are as follows. Firstly, what were the differences and similarities between the

NATO member nations? Secondly, what interests of the member nations were formed into common policies and measures? Thirdly, what were the internal and external pressures which may have ultimately determined the cohesiveness of the NATO alliance and hence the nature of collective decision-making?

The differences between the member nations of the NATO alliance have been well summarisied by Vandervanter, writing in 1963:

> Politically, NATO is composed of fifteen [now sixteen] independent and dissimilar nations. Their modes of government range from liberal democracy to benevolent oligarchy; their attitudes towards communism extend from complacent tolerance to belligerent antipathy. Their external interests, and hence some of their basic policies, differ with their size and location. Some members are world powers, whose global interests do not always coincide. Several smaller European states, on the other hand, with predominantly local oulooks, are primarily interested in preserving peace and . . . prosperity. Unless their own security is vitally and directly threatened, they would like to avoid involvement in any confrontation of the great powers.[3]

NATO is a voluntary alliance and has been termed a 'coalition of the willing'.[4] Its member states remained part of the alliance only by their own volition and were free to withdraw – or ignore collective decisions – at any time.

The purpose of the formation of NATO was, of course, establishing a common security policy in Western Europe and North America and making the consequent collective defence preparations to meet the goals of that policy.[5] The terms of the North Atlantic Treaty committed members to consultation (Article 4) and collective measures (Articles 5 and 6) to guarantee the territorial integrity of the treaty boundary areas. Central to this was Article 5 which states that:

> . . . an armed attack against one or more of them in Europe or North America shall be considered an attack against them all and consequently they agree that, if such an armed attack occurs each of them . . . will assist the Party or Parties so attacked by taking forthwith, individually and in concert with the other parties, such action as it deems necessary, including the use of armed force, to restore and maintain the security of the North Atlantic area.[6]

For reasons explored in Chapter One it was evident that the commitment to consultation was not as clear-cut as the treaty implied.

Furthermore the commitment to collective action was similarly ambiguous. NATO's own analysis of the North Atlantic Treaty stated that Article 5 '. . . leaves each signatory free to take whatever action it considers appropriate . . . to restore and maintain security'.[7] Since military action was but one measure by which security could be restored it was by no means certain that even in response to military aggression all NATO members would have agreed to collective military measures.

The pivotal question in the present context thus becomes: was it possible for NATO to build a political consensus around collective military action? In other words, to return to the title of this section, was coalition decision-making for collective military action achievable, or would national political decision-making have emerged as the principal determinant of military action?

This question cannot be answered with certainty, but it is possible from an assessment of the principal factors which were likely to play a part in consensus-building and from the available empirical evidence to approach some understanding of the possibilities.

The principal factors determining consensus-building were likely to be the need for political unity; the assessment of the nature of aggression based on information and intelligence; the perception of the gravity of the situation; the consultative process; and, the nature of the military response envisaged.

The need for political unity (to show resolve, unity of pupose, and to maximise deterrence) in the early stages of a crisis could have been the strongest cement holding the alliance together.

The first area of contention was likely to be the early assessment of the nature of aggression. Information and intelligence about the situation would probably have been confused, conflicting or ambiguous. The United States with by far the largest intelligence apparatus would inevitably have dominated this assessment. A number of problems would consequently have arisen. Firstly, some NATO allies may have had information which contradicted or differed from that of the United States. Secondly, the European allies with only limited intelligence of their own may have feared that the US was providing only selective information. Thirdly, the NATO allies would not necessarily have been sure that each was receiving the same information. Finally, at least some of the available information was likely to be ambiguous.

Despite NATO's formation of the Command Operations and Exercise Commitee (COEC) to attempt to establish a common assessment of the situation for the purposes of political decision-making it was argued in Chapter One that domestic and secrecy considerations would have limited the nature of information exchanged between allies. Partial, contradictory or confusing information would thus almost certainly have been circulating within the alliance.

It would have been in the interests of those seeking to build a consensus around military action to present a unified assessment of the situation perhaps leading to the provision of selective information or the disregarding of unhelpful, ambiguous or excessive information.

It was also clear that intelligence information was not shared equally amongst the NATO allies by the USA. The UK, for example, had a far closer intelligence relationship with the US than other allied nations.[8] This too could have contributed to the different perceptions of NATO member states.

The final point was the extent to which information provided, if ambiguous, could have led to different assessments by different nations within the alliance, each interpretation shaped by a host of nation-specific variables.

This last point leads to the second main problem associated with intelligence and information assessment which has to do with meaning and perception. Even if a consensus could have been reached around the situation assessment it does not follow that a consensus would necessarily have been reached around what the situation actually meant in terms of threat or potential threat to the NATO alliance.

The failure to translate threat evidence into political consensus of meaning – even within a single nation – has been at the heart of many surprise defeats and intelligence catastrophies in the post-war period.[9] The problem of reaching a similar political consensus in an alliance of 16 member nations was obviously even more complicated.

The third problem relates to the consultation process itself. NATO made considerable efforts to establish processes for, and to conduct, political consultation. The NAC or DPC provided the principal forum for the process whilst the crisis management system would have been instrumental in facilitating rapid information exchange and the General Political Guidelines were intended to provide the context for consultation both within the NATO structure and between national capitals. The limitations of these arrangements were explored in Chapter One.

Perhaps more important, however, was the nature of the consultation process itself. Whether consultation occurred prior to, during or after military steps had been taken, whether the concerns of all the

NATO allies would be taken into consideration, and whether consultation was to alter the perceptions and actions of individual nation states would largely have determined whether the consultation process could have been used successfully for consensus-building.

Finally, there was the question of reaching a consensus on the level of appropriate response. Each of NATO's member nations '. . . display[ed] the propensity of all states to seek control over their own military destiny'.[10] Any uncertainty would have acted against the prompt and open-ended commitment of NATO member's national military forces to a military response. Allied nations would have needed to consider a range of options – not only military – and the means and goals of any military response.

Even with NATO's advanced planning for information sharing, consultation and crisis management it is evident that there were a host of problems obstructing a prompt and clear political consensus on the response to crisis and military aggression.

Thus far we have drawn out structural difficulties in a theoretical way. But what does the empirical evidence of crises past tell us about whether these problems would have been likely to appear in practice?

There are at least eight crises of the post-war period which have some bearing on the present discussion: the Hungarian uprising of 1956; the Berlin crisis of 1961; the Cuban missile crisis of 1962; the Middle East war of 1967; the Czechoslovakian uprising of 1968; the Yom Kippur war of 1973; the Cyprus crisis of 1974; and the Polish crisis of 1980. It is instructive to look at political decision-making in NATO during each of these crises in turn (no attempt will be made to provide the context of these crises which the reader can find elsewhere).

At the time of the Hungarian uprising and subsequent Soviet suppression in November 1956 NATO was deeply divided by the Suez crisis of October 1956 which placed Britain and France at odds with the US. According to an analysis of NATO's response to the Hungarian crisis carried out for the Western European Union (WEU): 'any possible consultation within the NATO framework at the time of this [the Hungarian] crisis must have been rendered impossible by internal dissent following the Anglo-French invasion of Egypt [which precipitated the Suez crisis]'.[11]

In fact there was virtually no consultation, no political consensus, and no action taken by NATO in response to the Hungarian crisis because NATO was paralysed by disagreements between some parties by a dispute which arose outside the NATO framework.

The Berlin crisis of 1961 did lead to consultation between NATO allies in the North Atlantic Council but only after prompt and close discussions between the principal powers involved in the the crisis: France, the UK, the US and West Germany. Moreover, military preparations to meet the threat to Berlin were not taken within the NATO context (Western forces in Berlin were not considered part of NATO) but by a 'tripartite contingency planning staff [of France, the UK and USA] co-located with the NATO military headquarters'.[12] Throughout the crisis these three allies acted without consulting the other members of NATO, although the latter were briefed about events after they had taken place.[13]

The Cuban missile crisis occurred on the edge of the NATO theatre of operations (whose southern limit in the Atlantic is described by the Tropic of Cancer) but clearly could have brought NATO into confrontation with the Soviet Union within Europe. Despite the seriousness of this crisis the USA took military action involving a naval blockade of Cuba, the alerting of nuclear forces (including US nuclear forces deployed within the NATO theatre), and preparation for conventional combat all without calling a crisis meeting within the NATO framework.

Moreover, the US contact with NATO allies prior to the military decisions being taken was minimal. Just before the announcement of the decision to enforce the naval blockade President Kennedy sent former Secretary of State Dean Acheson across the Atlantic to explain his [Kennedy's] decision to the British, French and West German leaders.[14] Acheson is reported to have told the French President Charles de Gaulle that he was there to inform not to consult.[15]

NATO was informed through Acheson of Kennedy's decision[16] and the NATO Secretary General was reportedly informed that he had '. . . only a few minutes to decide whether to raise any objections'.[17] Throughout the remainder of the crisis NATO was primarily kept informed of events through US State Department briefings.[18]

The Arab-Israeli conflict of 1967 errupted on 5th June and NATO held an extraordinary NAC meeting to discuss the war on the same day.[19] NATO took no action other than the exchange of information and after the routine NAC ministerial meeting of 14th June (by which time the war had ended) issued a communiqué which merely noted the June 5th meeting.[20]

The Soviet invasion of Czechoslovakia in August 1968 virtually replicated the Hungarian suppression of 1956 but occurred at a time when NATO was not internally riven. In the context of the present discussion it was perhaps the most relevant of the post-war crises.

NATO had been monitoring the deterioration of Soviet-Czechoslovakia relations and the build up of Soviet forces on the border between the two during 1968 and made accurate assessments of force capabilities. But NATO failed to translate this into political warning and consensus by wrongly estimating that the Soviet Union would not risk an invasion for fear of the impact on East-West relations.[21]

The Czech crisis became notorious for the shortfalls it exposed in NATO crisis decision-making arrangements. Alliance leaders were notified of the invasion by Moscow and the news media, rather than by NATO.[22] In the words of one analyst: '. . . consultation and the flow of information broke down'.[23] Even more serious, however, is the following assessment of NATO's decision-making during the crisis:

> . . . when the NATO Council of Ministers met to consider alert measures after warning of the attack came in the early hours of the morning [the meeting did not take place until half a day later] the group debated each of the measures one-by-one and the deliberations gradually ground to a halt. Because approval by each of the [then] fifteen national representatives was required, not a single one was implemented.[24]

The Middle East or 'Yom Kippur' war of October 1973 occurred outside NATO's purview but had implications across the NATO theatre. At the height of the crisis US President Nixon and his Secretary of State Henry Kissinger ordered a world-wide alert of nuclear forces without consulting NATO allies. All US military commands around the world – including USEUCOM – were placed on DEFCON 3.[25]

Kissinger ordered the USEUCOM military commander (who was also SACEUR) to raise the alert level of US forces within the European theatre, but his only contact with NATO was to send an overnight telegram to the US political representative at NATO headquarters asking him to inform the allies of the alert.[26] The only other information provided by Kissinger was to inform the British Ambassador to the United States of the alert measure more than an hour after the event and to indicate to the latter that the US would brief the NAC more fully later in the day.[27]

The Cyprus crisis of 1974 was of a different complexion. The protagonists, Greece and Turkey, were both NATO members. While war on the border of the two countries was checked by the NATO

Secretary General, it was not possible to prevent a Turkish invasion of Cyprus on 20th July.[28]

In tripartite talks involving the UK, Greece and Turkey held in Geneva on 29th July Greece called, unsuccessfully, for an immediate Ministerial meeting of the NAC. NATO refused to convene such a meeting at short notice, despite the fact, technically, that Article 5 of the North Atlantic Treaty had been abbrogated. Greece responded, temporarily, by withdrawing from NATO's integrated military command and from the DPC.

The final crisis of relevance here is the Polish crisis of 1980. The deterioration of Soviet-Polish relations together with ominous Soviet military manoeuvres threatened a reprise of 1956 and 1968. NATO monitored the situation throughout the year but took no preparatory measures prefering instead to maintain a low profile in order not to risk the escalation of events. The NAC meeting of 11–12 December 1980 merely noted that '. . . the Soviet menace which hangs over Poland give[s] cause for grave concern to members of the Alliance'.[29]

Overall this empirical evidence seems to suggest a number of thematic problems for NATO crisis decision-making. Some of these have been usefully summarised by a WEU analysis of NATO crisis management covering the same crises:

i the perception of a crisis is subjective; the interpretation of events, which may well be ambiguous, is crucial;
ii the political assessment of intentions may well be wrong; it was in the case of the Soviet invasion of Czechoslovakia in 1968;
iii too frequently the political leadership fails to convert warning time into preparation time – military readiness measures are not ordered . . . ;
iv internal conflicts within the Alliance can endanger the determination of NATO to respond and act decisively.[30]

To these can be added firstly, the observation that certain allies – notably the US – have been willing to take military measures without alliance consultation even when this may threaten to embroil NATO in a military conflict. Secondly, a hierarchy of national actors within NATO was clearly evident in which the US, UK, France and West Germany emerged as undoubtedly more equal that other partners in crisis circumstances. Thirdly, the need for NATO to obtain political consensus on collective measures risked delaying decison-making and could have impeded decision-making altogether.

On this latter point a former SACEUR reportedly noted that '. . . the political process of decision could take days or weeks'.[31] The framework against which this time scale had to be measured was the warning time of Soviet attack which NATO had routinely anticipated in planning. This varied from as little as 48 hours to more conservative estimates of 8–14 days[32] or as long as a month.[33]

By contrast national decision-making processes of the member states were not hampered by the need to reach anything other than domestic executive consensus. These processes were more rapid, less complicated by conflicting perceptions, intelligence and analysis, and better able to reach timely political decisions.

The evidence of past crises indicates that national decision-makers have felt themselves prevented from consultating NATO by the need for timely decision-making. Writing later about the European anger at not being consulted prior to US military preparations during the Yom Kippur war, for example, Henry Kissinger noted: 'Abstractly, our allies were justified in their complaints. Realistically, we [the USA] had little time and we had to balance serious considerations.'[34]

Similarly during the Cuban missile crisis Kennedy ran into time difficulties, but also reportedly felt that he could not afford the luxury of seeking additional opinions from the NATO allies.[35] Kennedy clearly felt that further opinions could interfere with national decision-making.

During the Czech invasion of 1968 the US ordered limited military activity – the positioning of two US naval destroyers in the Black Sea – without allied consultation. The rationale given for the decision not to consult provides evidence of yet another factor acting against coalition decision-making. The then US Ambassador to NATO Harland Cleveland noted:

> When plans for the destroyers' visit were checked by the Navy with the State Department, it fell to me to suggest that we consult our allies about it ahead of time. The suggestion . . . was not well received: the clinching argument in Washington seemed to be the circular reasoning that if we consulted in NATO about the visit, it would be harder later on (when predictably the Soviets would complain we were being provocative) to allege that we regarded the visit as previously scheduled, normal and nothing to get excited about.[36]

For a variety of reasons, supported by empirical evidence, it seems likely that in a serious crisis coalition decision-making within NATO would have been hampered whilst national decision-making would have been relatively straightforward. Whilst efforts would undoubtedly have been made to build a political consensus around a given situation and response it was more likely that national measures would have been taken – particularly by the US and other leading players in NATO – and a consensus built later around that decision.

NATO, of course, accepted the force of this argument by providing a 'let-out' clause on commitments to consult which insisted on consultation only if time and circumstances permitted (see Chapter One).

The practical implications of this were that unless a clear political consensus could have been quickly built around the crisis situation based on a common intelligence picture, a shared view of the necessary response and the timely agreement of that response, national decision-making was likely to prevail. This could have meant that some nations within NATO would have begun military preparations whilst others moved more slowly down that path and still others rejected the military option outright.

The active nations within NATO could then have been faced with the task of cajolling reluctant nations into agreeing crucial collective decisions concerning NATO mobilisation including the alerting of NATO forces and the transfer of national forces to NATO operational command.

TRANSITION TO WAR

The process of transition to war did not begin with the Chop. Many mobilisation steps including calling up reinforcements, integrating communications and deploying forces to foward positions were intended to be made before military integration.[37] Obviously those nations most committed to military preparations would have begun these measures first with reluctant nations following later or perhaps not at all. The possibility that some NATO nations may have opted out at this early stage did not elude the United States. At least one source indicates that some American advanced planning had gone into covering military regions left undefended by the withdrawal or inactivity of some of the European allies. According to Richard Betts writing in 1982:

> Officers interviewed at Central Army Group and the US Eucom said they believed that with lead time of a week the Dutch or Belgian

> sectors [see Figure 2.3 in Chapter Two] could be covered adequately, but that this would leave virtually no operational reserve.[38]

The process of mobilisation for war would have required the alerting of forces. SACEUR and the other major MNCs had some authority to raise the LERTCON alert levels. Sources differ about whether SACEUR was authorised to raise LERTCON to the highest level but are agreed that above 'Military Vigilance' level SACEUR and the other MNCs were obliged to consult with the DPC or NAC.

The decision to raise the LERTCON for NATO forces consequently represented the first collective hurdle for NATO's political decision-makers. If their collective decision-making was delayed or piecemeal it would have been likely that national force alert levels would have been raised by those national governments committed to military preparations. Thus, for example, the US would have raised its DEFCON alert status as it did in 1962, 1973 and 1981.[39] Whilst forces could have been mobilised along national lines NATO could not have fought as a coherent whole without command integration. At some point either a military request would have been sought for command integration or a political directive would have been issued by the DPC (or NAC) on instruction from national capitals. There were, however, understood to be no NATO guidelines for when and under what circumstances this decision would need to be taken.[40] The possibility thus existed that political deliberation would have slowed the decision to chop forces whilst each nation decided whether the situation merited the step. The alternative to allowing this process to delay preparations was to proceed with national alert steps or even to begin a limited chop of those forces which were committed to military action. Since planning for the chop was premised on the participation of all the allies it is unclear whether NATO could have coped with integrating a partial force or whether NATO was capable of a staggered chop in which reluctant forces joined the command integration at a later date. It is important to note that NATO had never chopped its national forces, even in an exercise, let alone tried the command integration of partial forces. Any of these steps would have taken NATO into unknown territory.

To continue the analysis it is necessary to understand exactly what the chop entailed. The transfer of forces from national to NATO control meant that operationally the forces became subject to the direction of SACEUR (or the other MNCs) and subordinate commands acting on the instructions of the NAC itself informed by national capitals. The command chain for national forces would thus have been greatly complicated. The chop thus took national forces formally out of direct national control. However, national governments – as we have seen – would not have lost their ability to communicate with national forces or vice versa since strictly national lines of communications would have stayed open. National political leaders could thus have influenced decision-making indirectly through the collective process of NATO's political fora and directly through national systems. This duality was not without tensions.

Even after the chop national governments were at liberty to veto actions to which its national forces had been directed by NATO or to withdraw those forces from NATO control, notwithstanding that these steps could have created enormous military difficulties for the alliance.[41]

Moreover, while military action may have been strictly defined by NATO political leaders any deviation from the limits of action at the request of military leaders – such as SACEUR – may have seen national forces refering back to national political leaders for direction.[42]

THE DECISION TO USE NUCLEAR WEAPONS

This section deals with the first use of nuclear weapons by NATO in response to a deteriorating conventional battle in which NATO defeat seemed unavoidable, a scenario consistent with the flexible response strategy.

The decision to use nuclear weapons would have taken NATO across the nuclear threshold into a situation, to quote one source, 'with no clear stopping point this side of Armageddon'.[43] Moreover, it would have been a situation quite outside the experience of all concerned.

Nevertheless, the strategy of flexible response required the first use of nuclear weapons by NATO in response to imminent conventional defeat. The strategy specified that the use of nuclear weapons should be directed towards clear political goals. These included the signal of resolve and an indication of the intention to escalate still further should conflict continue, the signal that the use of nuclear forces had been limited and the desire to terminate the conflict at the lowest level consistent with avoiding military defeat and restoring the *status quo ante bellum*.

These were very precise goals and the ability of NATO's nuclear weapons to be used as so precise an instrument of policy is the subject of the final part of this chapter. For present purposes we need to try to assess the circumstances under which the decision to initiate the use of nuclear weapons might have been made and to ask whether this decision could have been made collectively, what scope existed for national autonomy of decision-making and how political leaders would have attempted to maintain the assured political control required by the strategy of flexible response.

To address the first of these issues it is safe to assume that the decision to initiate the use of nuclear weapons would only have been taken in the event of the imminent or apparently imminent defeat of NATO. Either the decision to use nuclear weapons would have emerged at the political level – collective or national – on the basis of the overall picture of combat political leaders had received or the decision would have originated with a request from military forces at whatever command level.

What is not clear was how political or military leaders would have judged the imminence of defeat and how far in advance of that defeat the decision to release nuclear weapons would have been taken.

The tensions within and between the political leadership of NATO at this stage would probably have been legion. Even assuming all member nations had agreed the alerting of military forces and the chop of national forces it was by no means certain that a collective decision to use nuclear weapons could have been reached. Assessments of the battle reaching national governments would probably have differed as indeed would the assessment of the necessity and wisdom of using nuclear forces.

This point was taken up by Ikle in his discussion of NATO's first use policy:

> . . . in an acute crisis, when nuclear destruction turns from an abstract fear into a concrete vision of terror, the . . . reliance on nuclear arms may make NATO governments afraid of their own military capability. Almost any step in military plans or operations that might bring the alliance closer to the threshold of nuclear war would be fiercely resisted by at least some governments. NATO's nuclear threat of 'first use' could turn inward to unravel the Alliance in the hour of crisis.[44]

By contrast Betts offered almost the opposite though to my mind less persuasive, argument:

> Perhaps in times of gravest crisis the political representatives, or the smaller countries, would react to strategic warnings by abdicating their prerogatives and handing decisions over to the "varsity" – the military respresentatives and the major countries of the Alliance.[45]

The General Political Guidelines agreed by NATO might have had a role to play in the process of reaching collective agreement. Intended to provide a framework within which the collective decision to use nuclear weapons could have been taken the GPGs might have provided the basis for collective agreement although they were not binding upon any of the NATO member nations. The GPGs had the advantage of a high degree of mutual agreement about their content (albeit agreement reached during peacetime) and, to the extent that the guidelines were prescriptive, could have provided a blueprint for action in a potentially chaotic situation. The attendant risk would have been that guidelines which prepared for the use of nuclear weapons and linked that use to politically desirable goals risked acting to hasten the use of nuclear weapons in a crisis or wartime situation.

To reach a collective decision to use nuclear weapons would have necessitated political consensus on a number of variables including the state of the conflict (and the prognosis), the alternative options, the potential risks of both using and not using the weapons, the precise political goals to which their use should be directed, and the nature timing and scale of use.

There is, of course, no empirical evidence upon which to assess whether it was possible for 16 sovereign states to have collectively agreed these types of decisions. Attempts to provide a clear answer to the central question 'can the use of nuclear weapons be decided upon collectively?' have in the past usually stumbled.[46]

Ultimately the decision to use nuclear weapons would have rested with the political leaders of the three nuclear powers – the US, UK and France. Each of these three could have initiated nuclear use independently. For the US and UK this was expressely provided for by the Athens guidelines. France was subject to no agreements on consultation. Nevertheless there would have been a strong political and military need to keep the Alliance together and thus a powerful motive for seeking to involve as many NATO allies in the decision as possible.

The point is thus whether these three nuclear allies could have agreed mutually to the use of nuclear weapons and built a consensus of support for the decision within NATO. With respect to nuclear weapons a clear hierarchy of states existed within the NATO alliance headed by the nuclear powers. The 1969 Provisional Political Guidleines (PPGs) gave extra weight to 'the views of the countries most directly affected – that is the country or countries on, or from, whose territory nuclear weapons would be employed; the country or countries providing the nuclear warheads; and the country or countries providing or manning the means of delivery'.[47]

To the three nuclear powers this would have added Western Germany and those other countries – Belgium, Greece, Italy, the Netherlands and Turkey – which participated in bilateral nuclear arrangements. The views of the other NATO nations would consequently have carried less weight.

A political consensus need not therefore have included all the allies, but only those directly involved with nuclear weapons operations whose forces would probably already have been engaged in combat.

However, any of the bilateral nuclear powers (that is those states which operated nuclear weapons only by virtue of their bilateral arrangements with the US) could have refused to allow their national forces to use nuclear weapons even if so directed by SACEUR or one of the other MNCs. Even if a political consensus to use nuclear weapons was reached it was still possible that certain nations would require that their forces were exempt from delivering the actual strikes. There would, from NATO's point of view, have been quite the reverse pressure to ensure that as many NATO nations as possible shared the nuclear decision and its execution.

In practical terms the US alone had the ability to override the decision of other governments to opt out of the nuclear decision and withold their bilateral forces, not by forcing the relevant nation or nations to comply but by covering the nuclear defence of the specific region or mission with its own preponderance of nuclear weapons and trained personnel.

The influence of the bilateral nuclear powers was thus limited to their veto over their own few forces and to their role in collective decision-making. At both levels they could readily have been overriden.

The first problem facing NATO would have been determining the state of the conflict. Even if NATO's command and control system had not been attacked this could still have proven a complex process. Different parts of the NATO theatre could have been in different states of balance. Forces in some areas may have reported imminent defeat while others may have achieved success. Different levels of military command may have had different assessments of the need for nuclear forces and the political and military leadership may also have differed. Furthermore information from national forces to national governments may have differed from that of NATO making the overall picture confused. If NATO's command and control infrastructure had been attacked this process would have been complicated further.

Secondly, the political leaders – advised from a variety of sources – would have needed to agree that the use of nuclear weapons was the option most likely to achieve success. If alternative measures offered the prospect of success without the attendant risk of nuclear escalation they surely would have appeared more attractive. Uncertainty could at least have delayed decision-making in the hope that the balance of the war could be clarified.

The question of agreeing the need to use nuclear weapons would obviously have been a critical decision for NATO. One of the strongest pressures bearing on the issue would have been the level of civilian casualties and collateral damage likely to be sustained in the use of nuclear weapons. The dense population of Germany and the Benelux countries, the most likely battleground in the period under consideration, made this a particularly acute pressure on these countries. As a consequence Bracken argued that: 'NATO cohesion, or at least resolve, may vanish in a crisis if a nuclear defense appears to be the only available option.'[48]

Thirdly, NATO would have needed to agree on the political aims of the use of the weapons. Whilst the range of the aims of the strategy of flexible response has been described it would still have fallen to political leaders to choose which of these aims to pursue. The intention to signal resolve might have argued for quite a different response to the intention to avoid military defeat. Compromising between the various aims to create an optimum solution might also have led to long deliberation.

Finally, NATO would have needed to decide on the nature, timing and scale of the appropriate nuclear response. To meet the political goals defined by its strategy NATO would probably have been concerned to use the most controllable nuclear weapons – that is weapons over which political leaders exercised a high degree of control. NATO would also have been keen to keep down the number of nuclear weapons used in order to convey the limited nature of the strikes to the adversary. Finally, depending on the political goals sought, the weapons would have needed to be directed against targets of particular relevance to the situation and to the signal intended to be sent by their use. Statements on NATO's intentions on this question were naturally infrequent and imprecise but former deputy SACEUR Admiral Sir James Eberle has stated:

> One could expect the first step on a nuclear ladder of escalation to be a small exchange of weapons. It would need to be politically relevant and it would need to be militarily credible, but it might be between one and two weapons or perhaps twenty or thirty, something of that order.[49]

The choice of the numbers of nuclear weapons to be used was an obvious point of contention amongst political leaders. The closest evidence available probably comes from NATO exercises – HILEX and WINTEX – which were specifically designed to test collective decision-making on the use of nuclear weapons and to exercise the GPGs. The last WINTEX exercise, held in April 1989, illustrated a common theme of such exercises which have consistently shown the US to favour larger nuclear weapons packages than their European allies. In the 1989 WINTEX the US argued for a strike of 21 weapons while the Europeans argued for substantially less and eventually compromised at 17 weapons. Later in the exercise the West German representative withdrew on the instruction of Chancellor Kohl.[50] Whether a similar pattern would have emerged in practice is less important than the evidence of deliberation which would have slowed the decison-making process.

The question of the controlability of nuclear weapons would also have been crucial for political leaders. For initial strikes to meet political goals required weapons whose targets, timing and accuracy could best be assured and over whose selection politicians could have exercised maximum control. This argued for weapons such as aircraft-delivered bombs and longer-range missiles rather than shorter-range systems – such as artillery – over which political leaders exercised only indirect control.

The question of the scale of use related not only to the numbers of weapons used but also to the size of weapons used and this in turn would have depended upon the targets to be struck. In the WINTEX exercise discussed above the Alliance agreed the use of 100kt weapons against a variety of targets in Eastern Europe and the Soviet Union.[51]

Finally, NATO would also have been concerned with the nature of use, trying to ensure minimum civilian casualties, minimum friendly force casualties and the military effectiveness of the strikes in the hope that these could have turned the tide of battle.

If the decision to use nuclear weapons was indeed taken because of imminent military defeat then the discussion process and the questions to be debated would need to have been agreed in a timeframe which would still have allowed the nuclear option to make a difference to the military situation. The question is whether it would have been realistic to achieve a political consensus on such critical questions under time pressure. If not collective decision-making on the use of nuclear weapons would have proved impractical. In this situation the nuclear powers would have made national decisions or attempted to co-ordinate their activities with a small group of allies, but NATO collectively would in practice have ceded any control over nuclear operations.

There were strong arguments for believing that once the nuclear threshold had been crossed the escalation of the use of nuclear weapons would have been difficult to prevent. But there were also counter-arguments that the first use of nuclear weapons need not have precipitated a holocaust providing the clear political objectives set out by flexible response were observed, indeed NATO's flexible response strategy was premised on the latter interpretation.

Admiral Eberle, for example, has said:

> I believe that there is a chance of having a firebreak after this one exchange, after you have crossed the nuclear threshold and let's say NATO fires half a dozen weapons over Warsaw Pact territory – the Russians would almost inevitably have to make a response, let us assume they make a response against Western Europe. At that moment I think there is the time when political leaders could get together and say look where really are we going? what is happening? because if we don't stop now there will be a danger of a total strategic exchange which could have the effect of destroying the world.[52]

THE SECOND USE OF NUCLEAR WEAPONS

NATO's strategy of flexible response required that NATO should be capable of escalating the use of force (or using the same level of force again) if previous action failed to meet military and political objectives. The strategy thus required that NATO was capable of using nuclear weapons a second (or indeed nth) time under political control.

Two main scenarios for second use can be envisaged. The first was one in which the initial use of nuclear weapons by NATO failed to meet NATO objectives, the adversary pressed his conventional advantage and NATO contemplated a second nuclear strike to again attempt to turn the tide of the conflict. The second scenario was one in which NATO contemplated second use after a Soviet strike of limited nature against NATO (if the strike were more general strategic systems would almost certainly have been engaged and attempts at escalation control would be virtually without meaning).

In the first scenario political leaders would have faced the same set of questions which had led them to agree the initial use of nuclear forces by NATO. These would, however, have been complicated by the failure of the initial strike to achieve its objective, the greater risk of retaliation and subsequent escalation, the level of destruction wrought by the weapons (perhaps including civilian and friendly force casualties), and the political problems created by their repeated use.

Even if time was found for collective agreement of the initial use it would not necessarily follow that additional time would have been available to discuss and agree second use. If first use was intended to halt imminent defeat and failed it is likely that very little consultative time would have remained before imminent defeat became a reality. Decision-making would almost inevitably have moved into national hands, with only limited consultation between principal allies possible.

In what was termed a 'departure from reality' the scenario of the 1989 WINTEX exercise was premised on an initial strike by NATO which did not meet with a retaliatory response by the Warsaw pact. Having failed in its initial strike NATO, in the exercise scenario, ordered a second strike of 25 nuclear weapons.[53]

The more plausible scenario, that NATO would have been contemplating the second use of nuclear weapons after a retaliatory strike by the Soviet Union raises even more complex issues. Firstly, it was impossible to predict the nature and scale of Soviet action. Some analysts of Soviet strategy have suggested that the Soviet Union rejected concepts of limited war and escalation control.[54] If correct the retaliatory large-scale use of nuclear weapons by the Soviet Union would effectively have ended any notions of limited war in Europe or elsewhere. Other analysts, however, have argued that the Soviets too would have had powerful motives for limiting the use of nuclear weapons and that consequently they could have been expected to have initiated use or responded to NATO first use in a limited way.[55] It is on the assumption of the latter – however dubious – that the discussion continues. In postulating second (or nth) use on a nuclear ladder of escalation NATO's strategy clearly did not anticipate a massive nuclear response by the Soviet Union to a limited nuclear strike by NATO.

Secondly, it was difficult to predict the targets against which Soviet strikes would have been directed, and hence to determine the status of NATO after the blow. Perhaps the strongest indication was provided by evidence of Soviet targeting priorities. This suggested the following list of potential priority targets:

i Strategic command and control sites
ii US SIOP forces
iii French and British strategic forces
iv Theatre Nuclear Forces capable of striking the Soviet Union
v US-NATO conventional power projection forces
vi US-NATO administrative/government sites
vii US-NATO economic/industrial sites[56]

On the assumption that the Soviet Union was seeking to limit conflict the strikes would perhaps have been directed with the aim of signaling limited intent and avoiding escalation to the strategic level. Strikes could have been confined to conventional targets to tip the tide of battle further in the Soviets' favour or against theatre nuclear forces and theatre nuclear command and control with the intention of limiting NATO's ability to execute subsequent theatre nuclear strikes.

NATO's own studies, however, conducted during the preparations of the Provisional Political Guidelines (see Chapter One) suggested that second use of nuclear weapons in selective strikes could have resulted in short-term advantage for NATO or 'quite possibly in a pause in the conflict'.[57] The studies explicitly argued that:

> . . . follow-on use should have the same purpose as initial use (to persuade the enemy to cease his agression and withdraw) and the nature of the use should therefore still be selective and designed to meet this political requirement.[58]

This assessment was fed into the PPGs and later into the GPGs and could have been expected to inform NATO thinking in crisis and war.

The problems of pressing on with a decision to use NATO's nuclear weapons a second time would have been manifold. NATO would have needed to obtain a clear picture of the extent of the damage inflicted by the Soviet Union which would almost certainly have included substantial casualties. It would then have needed to reach agreement that the use of nuclear weapons for a second time was the preferred option, have determined the political goals of second use, and then selected the targets and weapons for the strikes. For these discussions even to have taken place assumes that the Soviet strikes had not seriously degraded NATO's crisis decision-making organisation or communication systems linking political leaders.

NATO's planning for the follow-on use of nuclear weapons assumed that none of these problems would have presented unsurmountable barriers to second use by NATO.

To reach collective agreement on second use NATO would have needed to persuade the leaders of countries like West Germany and the Benelux countries – which would be likely to have sustained civilian casualties if not in NATO's initial strike then in Soviet retaliation – to sanction further casualties. Moreover, since NATO's own studies showed only temporary advantage to NATO in a second strike and none if the Soviets retaliated a second time, NATO would have required these nations to agree second use with no clear expectation of success. At some point for many European allies the fear and cost of defeat would have been outweighed by the cost of continuing to prosecute a defence. If that point had not been previously reached the second use of nuclear weapons could have proven an intractable obstacle.

An additional concern would have been the French. This book has argued that France had the means and could have had the motive to co-ordinate an initial nuclear strike with NATO. There was, however, no provision in French strategy for follow-on use or graduated escalation. For second use by NATO to be limited would have required either that the French refrained from using its weapons or that France attempted to co-ordinate a second limited strike if such an option were possible. The French could have reasoned that the limited nature of NATO's initial strike had failed to meet the aim of France's pre-strategic planning, namely to signal the imminence of the use of French strategic forces, and have gone their own way.

One senior military figure has expressed the opinion that European participation in collective NATO defence would have evaporated after initial use of nuclear weapons.[59] The logic of his argument was that follow-on use was dubious being hard to control (for reasons considered below) and difficult to distinguish from war-fighting strategies. The Europeans could at best have been expected to agree the use of nuclear weapons once – though this would have been fraught with attendant risks and dangers – but could not have been expected to risk the decision a second time.

Assuming for the moment that this assessment is incorrect and NATO allies could have agreed the need for second use at the political level, NATO would again have faced questions concerning the weapons, scale and timing of use. The GPGs suggested that a similar scale of use to initial use was envisaged by NATO against similar targets and with similar weapons in order to restate the message.

THE POLITICAL-MILITARY INTERFACE: DILEMMAS AND TENSIONS

A range of military issues would undoubtedly have exerted an influence on any political decision-making in NATO about the preparations for and authorisation of the use of nuclear weapons. Conversely, political decision-making would greatly have shaped the nature of military actions governing the use of nuclear weapons. It is consequently important to understand the effect dilemmas and tensions at the interface of political and military decision-making may have had on the command and control of nuclear forces in NATO.

The tensions between political and military leaders flow from differing perceptions of aims and the means to achieve those aims in both crisis and war. Not unnaturally NATO's civilian leaders would have seen crises and war in political terms and have intended actions – including the use of force – to send political signals to the adversary.[60] The strategy of flexible response made the goals of these signals quite explicit and therefore placed strict limits and imposed tight political control on military action.

By contrast the military would probably have wished to take a range of preparatory steps during crises designed to minimise the prospects for adversarial surprise attack and to maximise their ability to achieve key military goals – if necessary by the escalation of conflict – in the event of war. Tensions may have arisen because, for a number of reasons, military and political goals were not always compatible and may even have been in direct opposition.

The first tension arose around the question of preparedness versus provocation. A number of military measures, some of which are explored below, would have been necessary to prepare for action whether offensive or defensive in nature. In heavily militarised Europe difficulties arose precisely because almost any military action could have appeared provocative, risking the very escalation of crises politicians may have been seeking to avoid. Defensive measures may not have been interpreted as such by the adversary. Moreover, the steps taken to prepare for action could themselves have signalled the inevitability or imminence of war.[61] Thus political leaders may have been keen to ensure that military provocation was minimised even if this interfered with military preparedness.

This point was particularly apposite at the nuclear level. To ensure its nuclear response capability NATO needed to begin preparations for the use of nuclear weapons early in a crisis despite the political arguments against it.

A second tension arose with respect to the level of military action considered appropriate. In his seminal study of the political-military interface Betts notes that the evidence of past wars and crises suggests that although soldiers were no more likely than civilians to recommend the use of force in a given situation, there were clear differences about the level of force thought necessary once the decision to use force had been made:

> Generals prefer using force quickly, massively and decisively to destroy enemy capbilities rather than rationing it gradually to coax the enemy to change his intentions.[62]

As a result military forces within NATO would probably have created upward pressures on the scale and nature of force – including at the nuclear level – and continued to sustain this pressure even after initial use. The typical planning size of a single Corps nuclear weapons package, usually between 50 and 100 weapons and therefore far larger than political conceptions of limited use, provides evidence to support this assessment.[63]

A third source of tension flowed from this. It was the military concern that 'political leaders concerned with national objectives [would] make victory difficult, more costly or even unattainable by putting restraints on the use of force'.[64] Military leaders would probably have resisted political management to the extent that they felt it was interefering with their ability to achieve desired military goals.

A fourth source of tension was that military objectives and perceptions may have created pressures not only for vertical escalation in requesting greater firepower but also horizontal escalation in seeking increased firepower at a greater number of locations across the theatre than political leaders may have prefered to sanction.

A fifth source of tension arose from the lack of civilian knowledge about military matters, particularly concerning what military measures actually implied in terms of activities, what was and was not achievable in military terms, and what was the best way to prosecute a conflict.[65]

On the first point there have, for example, been a number of occasions in which political leaders have authorised military alerts unaware of the full military implications of their actions. One such occurred in 1960 when the then US Secretary of Defense Thomas Gates ordered '. . . a quiet increase in military preparedness . . . without public notice if possible'.[66] Within a few hours this resulted in the US JCS placing the forces of each of the Unified and Specified command (including USEUCOM) on DEFCON 3 alert. Gates later admitted '. . . some commands went further in executing the instructions issued by the JCS . . . than I had anticipated'.[67] Gates' problems arose from his lack of understanding of the military implications of raising the DEFCON and his consequent imprecision in issuing orders which allowed the military to overstep his intentions.

On a second occasion in 1981 US Secretary of Defence Caspar Weinberger raised the alert status after the attempted assassination of President Reagan unaware of the military implications of his action. The then Secretary of State (and former SACEUR) Alexander Haig questioned Weinberger about his actions and later wrote 'I began to suspect that Weinberger did not know whether he had raised the DEFCON or, if he had raised it, to what level'.[68] Haig's suspicions were well founded and Weinberger was forced to check with the military to find out what he (Weinberger) had in fact done.

Events like these illustrate a key problem at the political-military interface:

> politicians will approve military actions without fully understanding their scope, their meaning, or how threatening they might appear to the adversary. Civilian leaders rarely have a detailed grasp of military planning, of alert procedures, or of rules of engagement, and might authorise action, only to later find that these actions jeopardise their control over events.[69]

This problem appears perhaps to run counter to the point made above that political leaders will seek to minimise escalatory signals, but both in fact illustrate the wider point that political leaders may seek to control military operations with insufficient knowledge to do so. This may in the case of NATO have resulted in necessary military preparedness being delayed or circumscribed or in an effective loss of political control as actions were authorised which exceed political intentions. As a result political leaders may have found themselves increasingly dependent upon the advice of their military leaders with clear implications for the shift of real decision-making power within the Alliance in crisis and war.

A sixth source of tension concerned the military need for timely political decisions in response to changing conflict dynamics. If political decision-making could not keep pace with military events there would have been a tendency firstly for collective political decision-making to break down and secondly for decision-making to devolve to the military level. The latter is one of the main reasons why political control of battlefield nuclear weapons was limited to the selection of parameters for SEPs (see Chapter Two). To the extent that political leaders resisted this downward pull of decision-making they would have created tensions and perhaps practical problems for the military.

A second thematic problem at the political-military interface flowed from the autonomy of military action. As has already been discussed SACEUR and the other MNCs had the authority to raise the alert level at least one rung (to Military Vigilance) without political consultation. On at least one occasion, during the 1968 Czechoslovakian crisis, this step was taken autonomously by the military despite the fact that NATO's political leaders '. . . purposely refrained from precautionary changes in military posture' in order not to risk the escalation of the situation.[70]

This event touches on the heart of the problem of military autonomy which was that action by the military risked sending unwanted political signals to the adversary and thus risk the escalation of a crisis. The problem arose because it was not practical to fully circumscribe military activity from the political level. All military forces operated standard operating procedures (or rules of engagement) which governed the limits of their activities without additional political or military orders. Within NATO, because of the 'enormous variety of weapons and weapons platforms, the various threats they face, and the different political environments in which contact with an adversary may occur . . . rules of engagement [were] highly complex'.[71] Political understanding of these rules – and of consequent permissable military action – may have been limited. During the Cuban missile crisis, for example, SOPs for both US surface naval vessels and for reconnaisance aircraft half way round the world threatened the stability of the crisis at critical moments. In the case of the former, US naval vessels pursued Soviet submarines with a vigour which 'violated the spirit of the Executive Committee [i.e. political] policy'.[72] This included the use of depth charges which reportedly brought several Soviet submarines to the surface.[73]

The latter concerned a US U-2 aircraft which strayed into Soviet airspace whilst on a routine mission at the height of the crisis. US fighter aircraft scrambled to assist the U-2, a decision taken by local commanders. News of the event caused great anxiety amongst America's political leaders. Secretary of State Robert McNamara reportedly exclaimed '. . . this means war with the Soviet Union'.[74] President Kennedy responded to the news by declaring: 'there is always some son-of-a-bitch who doesn't get the word', an assessment with central relevance here.[75]

Similarly, if less dramatically, the NATO alert ordered by SACEUR during the Czechoslovakian crisis led to some military actions which were deemed provocative by political leaders. Some British aircraft moved to dispersal airfields as a precautionary measure were, for example, ordered back to their original bases because of the provocative signals the movement could have sent.[76] The point to note is that even in a crisis political control of military action could not have been guaranteed.

A decision to alert military forces raised a number of political-military tensions. Delays by national political leaders in authorising the alert of their respective forces could have taken the different national forces within NATO out of alert synchronisation, meaning some forces would be closer to readiness than others. The process of decision-making or divisions within the NAC (or DPC) could have delayed the authorisation of higher LERTCON level – a problem which would have deepened as the military forces advanced up the LERTCON levels towards war. This risked making it very difficult for NATO to synchronise theatre-wide operations and thereby risked creating military vulnerabilities.

The alternative may have been that SACEUR and the other MNCs would have pressed on with military preparations without clear political authorisation from NATO. National military forces may consequently have proceeded with alert preparations in broad step with SACEUR or other MNC instructions even without clear national political authorisation. This would clearly have raised additional questions about military autonomy. Moreover, once begun, the preplanned military steps for mobilisation and combat readiness could have created a momentum driving military activities. Difficulties would be underlined if adversarial activity was also creating time pressure.

The alerting of forces implied not only changes in force deployments and readiness but also potentially in standard operating procedures and the delegation of authority.[77]

Political leaders may not have been fully aware of the nature of events they were authorising by agreeing to raise alert levels and may as a consequence have lost control over the political signals being sent by military activity. Part of the reason for this was that the process of alerting – to try to ensure smooth transition to war and to optimise response to warning of conflict – was largely preplanned. The political decision to raise the LERTCON level from 4 to 3, for example, directly implied political agreement to a range of military measures which may not have been fully understood at the political level. There would have been only limited scope for political leaders to attempt to shape or circumscribe alert measures which would have been occuring across the NATO theatre. Even limited political 'micro-management' of military preparations would have created problems if the instructions interfered with activities the military deemed necessary.

Within the context of the mobilisation of NATO forces initiated by the raising of alert levels the alerting of nuclear forces represented a crucially important element. Within NATO a consequence of raised alert levels would have been that pressures would have arisen for the placing of nuclear forces at raised alert. This has been made explicit for US forces, but similar pressures would surely have obtained on the other nuclear powers and the bilateral nuclear forces:

> A central tendency of the American security establishment is to establish a hedge against escalation by backing up regional conventional forces with a credible and ready nuclear force. To this end, nuclear alert steps that reduce critical vulnerabilities in the peacetime nuclear posture would be taken. Military prudence would recommend taking those steps early in a crisis while there was still time.[78]

The need to reduce critical vulnerabilities of the nuclear system meant that substantial measures needed to be taken to improve the readiness and protection of nuclear forces at a relatively low threshold of alert. For US forces DEFCON 3 – and because of the close link between the two for NATO forces LERTCON 3 – would probably have been the level at which important nuclear alert measures would have been taken (see below for a discussion of the escalatory potential of this). This was just one rung above the level of alert which military leaders were authorised to take themselves. Poilitical leaders could thus have been in the position that the first level of alert they had to authorise also agreed initial preparations for the use of nuclear weapons. Moreover, the military prudence of early alerting of nuclear forces during crises would probably have placed political leaders under considerable time pressure to agree the step.

The alternative was the deliberate witholding of political authority to alert nuclear forces, or the selective agreement to alert only a limited number or specific types of weapons within the theatre. This would have had the advantage of sending a political signal of NATO's limited intentions. This decision would, however, have risked tying military hands early in the crisis, and perhaps created vulnerabilities within NATO's nuclear forces and limits on NATO's subsequent nuclear options.

A final point about alerting, particularly of nuclear forces, was the notion of mutually reinforcing alerts between NATO and the Warsaw Pact in which an alert by one side was matched by a counter alert on the other, beginning an escalatory spiral of alert and counter-alert with potentially uncontrollable consequences. To the extent that this analysis, expounded by Bracken[79], had credence at high political levels it may have exerted an influence on decision-making in that fear of the upward spiral of escalation may have deterred political leaders from taking the decision to raise the alert status of nuclear forces.

Nowhere was the possibility for political-military tensions greater than in the decision to disperse NATO's nuclear forces during a crisis or perhaps at the onset of conventional war. The problem was both one of preparedness versus provocation and of preparedness versus the risk of the loss of political control.

The majority of nuclear forces at NATO's disposal were not maintained at operational readiness. Most were in the controlled environments of storage facilities.[80] While the weapons remained in storage facilities they offered a relatively small set of lucrative targets to a potential adversary.[81] Richard Perle, former Assistant US Secretary of Defense, has consequently argued:

> . . . it is vital to the execution of NATO strategy that the [nuclear] weapons would be dispersed well in advance of a Soviet attack . . . Once the weapons are dispersed, they present not a handfull of targets but many hundreds of targets, and the Soviet capacity to destroy all of those would be very doubtful indeed.[82]

Moreover, the time taken to remove the weapons from storage and match them to operational delivery systems could have created problems of preparedness for NATO if the decision was not taken relatively early in a crisis.

There were thus at least two powerful military reasons for the early dispersal of nuclear weapons from storage sites in a crisis, an event incidentally which Gottfried and Blair argued would be likely to have occurred at LERTCON 2.[83]

By contrast there were powerful political motives for resisting this decision. The first was that the sudden – and highly visible – dispersal of nuclear weapons to perhaps hundreds of sites across NATO would have sent a strong political signal to the Soviet Union that the use of nuclear weapons by NATO was possible and perhaps imminent. This could, of course, have exerted a strong deterrent influence, but it could equally have escalated the crisis irretrievably. Political leaders intent on crisis management and the avoidance of war might have considered it too dangerous a step to take.

The second was that the dispersal of nuclear forces across the battlefield into areas of potentially intense combat risked the loss of political control of the weapons due to the disruption or breakdown of command and control systems linking weapons to high level authority.[84]

Political leaders thus probably would have been deeply reluctant to heed military calls for the dispersal of the weapons. Moreover, political dissent was likely to have been greatest amongst countries like Western Germany on whose territory early dispersal would most likely have taken place. The fact that dispersal of nuclear weapons was envisaged at LERTCON 2 may also have impeded political agreement to raise NATO's alert level unless some form of political selectivity could be applied to the alert measures.

Directly and closely related to the question of the dispersal of nuclear forces was the question of the release of nuclear weapons. For some of NATO's TNF – those at least that needed to be matched to delivery systems from storage – the decision to disperse nuclear weapons to perhaps many hundreds of sites across the NATO theatre raised a fundamental military problem. Since it was necessary for political leaders to authorise the release of nuclear forces by providing use authorisation and release or PAL codes in order for weapons to function and since there were considerable problems with assured command and control after nuclear forces have been dispersed into the battle area there would probably have been powerful motives to provide nuclear release codes at the same time as weapons were dispersed from storage facilities.[85] From a political perspective this risked the loss of political control over the weapons and the loss of escalation control.

Any political resistance to the dispersal of nuclear weapons would have been heightened further if there was a simultaneous or implicitly linked request that dispersal be accompanied by nuclear release. Since NATO had to assume that the Soviets were able to monitor the release decision the political signals sent by the action would be even more hostile and threatening.[86] Linkage of the two – dispersal and release – risked not only hardening political resistance to early dispersal and complicating decision-making it may also have underlined differences between many of NATO's political leaders making collective agreement harder to reach and deepening political-military tensions.

It is possible that NATO would in practice have refused to release any of the nuclear weapons in storage precisely to avoid these problems and instead rely on nuclear systems such as aircraft and longer-range missiles which would have been outside the immediate battle area and subject to tighter political control. Such a decision would, however, have abrogated NATO's declaratory strategy (by greatly limiting the flexibility of flexible response), and rendered large numbers of NATO's theatre nuclear weapons unusable.

NUCLEAR WEAPONS COMMAND AND CONTROL

In concentrating on the political dimension of nuclear decision-making the foregoing has made many implicit assumptions about NATO's nuclear forces. The most important of these are that multinational multilingual forces could have operated nuclear weapons collectively, that NATO's nuclear command and control infrastructure was sufficiently robust to allow weapons to be politically controlled during crisis and war, and that NATO's command and control infrastructure was capable of directing nuclear weapons to meet political aims. This final section provides evidence that these assumptions were flawed and that political decision-making would have been hampered further by limitations in the performance and robustness of NATO's nuclear command and control infrastructure.

Three issues dominate these concerns: interoperability; technical limitations and vulnerabilities; and system performance in crisis and war.

To deal with each in turn: interoperability was central to the ability of NATO's disparate multinational forces to fight as a cohesive whole. According to NATO:

> interoperability is a minimum requirement for military effectiveness since it enables systems, units or forces to provide services to, and accept services from, other systems, units or forces and to use the services exchanged in a manner which enables them to operate effectively together. The need to be able to intercommunicate, for example, is self-evident.[87]

With respect to command and control systems within the theatre NATO took one of two approaches to the procurement and operation of systems. Either it funded a NATO-wide and NATO-owned system which had structural interoperability across the theatre (the obvious example would be the NICS programme discussed in Chapter One) or it permitted the development of national systems which it then sought to integrate and make interoperable in one of two ways. The first was to encourage the standardisation of equipment deployed by the various national elements within NATO. The second was to develop interface technology which allowed different command and control networks to be linked together.

The evolutionary nature of NATO's command and control system, the development of national systems, the lack of overall system architecture, and budget limitations, all contributed to the interoperability problems which faced NATO during the period of this study.[88]

Within NATO it was possible to identify at least six levels of interoperability interfaces: the national to NATO political interface; the NATO political to MNC; the MNC to Corps level (or Air/Sea equivalent); the Corps to field soldiers; NATO to France; and national interservice interoperability. While it was not necessary (or practical) to design systems which allowed each of these levels to communicate and interoperate with each other it was the case that the less interoperability existed the longer and more complex the pathway by which orders or information had to be passed up and down the command chain and between parallel command chains.

The issue was complicated further by the need for horizontal interoperability between air, sea and land forces within NATO; by the need for interoperability between new and older technology; and the need for interoperability in a variety of systems including communications, CCIS, ACCS, air defence, attack warning and characterisation, and intelligence.

Between national capitals and NATO headquarters communications were secure, reliable and interoperable following the development of NICS. Prior to this, throughout the 1960s and 1970s, communications at this level were less straightforward being largely dependent upon the technically limited TARE system (see Chapter One). Intelligence was shared on the basis (and with the problems) outlined in Chapter One. CCIS are not relevant at this level.

Between NATO Headquarters, SHAPE and the principal MNCs communications were also secure, reliable and interoperable being again served by NICS. Prior to NICS communications at this level were dominated by the ACE-HIGH network which offered interoperability but only limited technical capability. CCIS developed separately within NATO headquarters, SHAPE and the MNCs and there was consequently no interoperability of these systems.[89] Battlefield intelligence was based on national collection with only limited means for synthesising information at command centres. The implications of this for a clear NATO assessment of any particular battle area should not be lost.

Within each MNC at least down through PSC to Corps level communications were based on NICS and were therefore fully interoperable. Prior to this communications were dependent upon ACE-HIGH and national systems with no interoperability amongst the latter (other than commercial telephone networks and carriers based on commercial networks). CCIS were being developed at each command headquarters on an *ad hoc* basis to meet individual command centre needs. However, through the purchase of commercial equipment or through single contracts for CCIS within each MNC or PSC region a reasonable degree of interoperability was achieved by the late 1980s.[90] There was, nevertheless, virtually no horizontal CCIS interoperability between PSCs or Corps in different MNCs.

Below Corps level the problems of interoperability were legion. NATO battlefield communications networks included at least six systems: Ptarmigan (UK), RITA (France, Belgium and the US); AUTOKONETZ (West Germany), ZODIAC (Netherlands), MSE (US) and Tri-Tac (US). None of these systems were directly interoperable with NICS and there were interoperability problems between them.[91] A US Congressional study of ACE communications in 1988 concluded: 'seven NATO countries are developing six different battlefield communications systems, none of which can talk to the others'.[92] Italian, Greek and Turkish battlefield systems fared no better.

As a result SACEUR could not directly reach any of the national army corps under his command and, according to one author, therefore could '. . . not accomplish the major aims of a modern C^3 system'.[93] These problems of interoperability created considerable difficulties of sideways communications between fighting units (at Corps, Division or Brigade level) of NATO's multinational forces.

The introduction of both CCIS and intelligence and target acquisition systems in the battlefield were also carried out on the basis of national systems (see Chapter Two). There were consequently parallel interoperability problems in these missions areas to those in communications.

The effect of the lack of interoperability between these systems was to compartmentalise national forces below Corps level and greatly limit the prospect for horizontal co-ordination during battle implying that operational control would have followed national lines and that co-ordination could have been highly problematic, particularly in a hostile battle area.

To the extent that interoperability problems undermined NATO's ability to fight a conventional defence successfully they could have hastened the use of NATO nuclear forces and thereby been instrumental in the escalation of conflict to the nuclear level.

Interoperability problems could also have created difficulties for the co-ordination of nuclear operations at a number of levels: firstly, in those areas where national nuclear forces conducting air sea or land operations overlapped; secondly, in those areas where combined nuclear operations needed to be co-ordinated (for example, two adjacent Corps co-ordinating nuclear strikes); and, thirdly, in operations where the nuclear forces of one nation within NATO were used in the defence of the forces of another.

The result may have been to compartmentalise nuclear operations along national lines below Corps level; reduce the effectiveness of combined operations; and, complicate the process of the effective selective release and use of nuclear weapons.

From the discussion in Chapter Two it was evident that NATO's command and control system evolved in an *ad hoc* way in response to changing strategies, weapons technologies and command and control technologies, both of NATO and the Warsaw Pact.

Systems differed in age and consequently not only in the nature of technology deployed but also, crucially, in the mission which they were originally designed to serve. NATO's ACE-HIGH network, for example, dated from the late 1950s (though not fully operational until 1963) and was intended to deliver a one-way message from SHAPE command headquarters (then based in France) to all ACE commanders authorising, *inter alia*, the general release of nuclear weapons consistent with NATO's strategy, at the time, of trip-wire retaliation.

Although upgraded ACE-HIGH remained an inflexible system quite unsuited to the command and control missions implied by the strategy of flexible response. ACE-HIGH operated close to, but not actually in, real time; it was not designed for survivability in war; sites could readily be severed by striking a limited number of points in the network; and its ageing technology was prone to failure.[94]

This analysis serves to illustrate the wider point that both national and NATO command and control systems designed or deployed in Europe before the change from a trip-wire strategy were unlikely to be appropriate for flexible response missions. Such systems include AUTODIN, AUTOVON, NARRS, and elements of the UK's DCN.

Problems amongst more modern systems included the lack of funding for NATO command and control systems, the piecemeal nature of system design and development, the compromise of system specification, and the problems of providing system security.

The piecemeal and evolutionary development of national command and control systems created a heterogeneous and unplanned system architecture in NATO in which deployed systems offered different performance capabilities, different reliability, different system hardness against attack, and different systems security against monitoring and electronic warfare. The differences would probably have become most evident during combat and would have influenced NATO's performance as a cohesive military whole. Nations – particularly the US and France – that invested heavily in relatively secure, survivable, and reliable command and control systems, particularly for essential missions such as the use of nuclear weapons, risked finding the less capable command and control systems of their allies (or of NATO) degrading disproportionately in an escalating military situation.

Technical assessments of the overall capabilities of NATO's command and control system were made by a number of informed observers. General Richard C. Ellis, former CINC of Strategic Air Command noted in 1982: 'if you think we're in trouble in strategic connectivity, we're in terrible trouble in the battlefield theatre area'.[95]

In his authoratative study of theatre communications and command and control information systems, John Cushman wrote:

> For the typical senior commander, allied or US, whose forces must use these systems, they represent the largely unplanned splicing together of ill-fitting components which have been delivered to his forces by relatively independent parties far away who have co-ordinated adequately neither with him and his staff nor with each other.[96]

Perhaps one of the most telling assessments, however, was that made by Robert Raggett, the Editor of *Jane's Military Communications* in 1984:

> NATO command and control is currently in a state of some disarray. Political problems have thrown off course any overall strategy that might have existed, while the inability of industry to produce and deliver equipment and systems for NATO C^3 programmes has seriously reduced overall C^3 effectiveness. Some even argue that with the present unco-ordinated mix of old and new technology, the lack of a totally structured system architecture, and the delayed state of many programmes, in the event of a major armed conflict within Europe NATO C^3 could evaporate into a state of total ineffectiveness and confusion.[97]

Kelleher's 1987 analysis of NATO command and control restated the point:

> The present system of command in NATO, especially in the thickly settled central front, is clearly less than satisfactory. Should war come, most observers forsee chaos, even without initial TNF use.[98]

To meet the operational requirements of flexible response NATO's command and control system needed to support military operations in crisis and conventional and nuclear war. NATO expected its command and control assets to come under direct attack in the early stages of conflict. The means by which these attacks could have been carried out were manifold but the most important included: sabotage; electronic warfare; conventional attacks; nuclear weapons electromagnetic effects (EMP and TREE); and direct nuclear attack.[99] Large sections of NATO command and control infrastructure was vulnerable to these threats and virtually all NATO command and control targets were vulnerable to limited nuclear strikes. To significantly degrade or even destroy NATO's command and control apparatus would not have required large-scale nuclear strikes.

Ball identified 60 targets comprised of command centres, early warning sites, intelligence stations and communications centres which he argued 'if destroyed or incapacitated, would completely nullify any capacity for controlled, informed or responsive military operations within the theatre'.[100] Furthermore, and perhaps more tellingly, he argued that 'it is likely that the destruction or incapacitation of less than 20 of these 60 sites would totally immobilise command and control across the whole European theatre'.[101]

Attempts to degrade NATO's ability to respond with theatre nuclear forces could have targetted the two command centres at SHAPE and USEUCOM (High Wycombe) together with those at Norwood (UK), High Wycombe (RAF Strike Command), Taverny and Val d'Oise (France). Pirmasens in Germany as the headquarters of NATO's PAL teams, the ECCCS and the Cemetry and Regency Net systems would also have been an obvious and lucrative target. The destruction of these targets could have left political leaderships and nuclear forces in place but greatly complicated the control of the latter by the former.

The destruction of just the MNC command headquarters within the European theatre (SACEUR and CINCHAN) could have created problems of command and control at the political-military interface, greatly complicated coalition operations and shifted the emphasis to national operations, highlighting national differences within the alliance.

Attacks against Corps-level headquarters would have exploited the command and control problems at the interface between NATO theatre systems and national command and control systems deployed below Corps level making cohesive action by NATO virtually impossible thereby further underlining reliance on national lines of command and control.

Attacks against command and control systems which linked the US to the European theatre – including the WWMCCS, Autodin and Autovon, NARS, Giant Talk, and DSCS ground entry points could have underlined the theatre nature of the conflict, and symbolically at least separated the fortunes of the continental US from those of Europe.

What was evident from any assessment of command and control vulnerability is the idea that not all NATO member nations were equal in command and control terms. Even in large scale theatre nuclear strikes much of the US's command and control apparatus would have remained intact simply by virtue of being outside the theatre. US command centres, strategic forces, satellites, satellite control stations, early warning and intelligence assets would all have had a greater endurance than comparable European systems. Even within the theatre the US's nuclear command and control apparatus was more hardened and robust than its European counterparts, suggesting, after a certain point in a conflict, that the US (and/or possibly the other nuclear powers) would have conducted nuclear operations, for all practical purposes, alone.

The strategy of flexible response required that NATO should be willing and able to initiate the use of nuclear weapons in a controlled and selective manner. In practical terms this required endurance from NATO's nuclear forces and nuclear command and control systems in order that NATO could call upon specific nuclear weapons at any stage of a conflict.

Controlled escalation in NATO would have been dependent upon NATO (or the nuclear powers) having a relatively clear and close to real-time view of what was taking place across the theatre. Yet in a fluid conflict with NATO's command and control system under widespread attack, perhaps for a considerable period, it must be questionable whether NATO would have been so well informed. It was precisely those command and control assets of NATO necessary for the control of nuclear operations which were most likely to come under attack at the outset of, and during, combat. With degraded sensors, communications and decision-making it strains credibility that NATO could have assessed the progress of battle accurately, made choices about first, second, or nth use to support NATO's aims, then identified nuclear forces which had survived, were still under political control and were still relevant in the sense that they could be brought to bear on the chosen targets within a relevant timespan and in the controlled and limited way required.

A complicating factor and the central dilemma of nuclear war termination in the European theatre was that the command and control assets which were needed to terminate nuclear exchanges and maintain the ceasefire were also the assets which would have come under attack if an adversary were seeking to degrade NATO's nuclear capability.

The task for NATO would have been to re-establish communications with its nuclear weapons across the theatre and impose high-level control on the weapons. For a number of reasons this would have proven problematic.

Studies of nuclear war termination have tended to concentrate on strategic nuclear systems[102] but these provide some guidelines for an analysis of theatre nuclear war termination.

The first point to make is that NATO's command and control system was not designed with the specific goal of nuclear war termination in mind. Older parts of the system (such as the ACE-HIGH) were in fact designed for exactly the opposite purpose: the top-down, general and open-ended release of nuclear weapons consistent with the trip-wire strategy which predated flexible response. More up-to-date systems sought to build in robustness and survivability for controlled options but even these were vulnerable to a variety of threats and consequently subject to degradation and destruction. Moreover, there was virtually no provision for the regeneration of command and control networks.

A system designed for nuclear war termination would be predicated on maintaining firm political control of all nuclear weapons at all times or the ability to re-establish lost connectivity. Yet within the NATO theatre this cannot be said to have been the case. US naval and British nuclear weapons were not controlled by PAL technology placing the burden of political control of these weapons on military obedience. The lack of assured and survivable command and control risked the widespread loss of direct political control of nuclear forces. Finally, the emphasis on the early dispersal of large numbers of battlefield nuclear weapons – whether released simultaneously or not – created, at least according to many analysts, the risk of loss of political control.

Since it is axiomatic that nuclear weapons would have been used if NATO was attempting to terminate a nuclear war it would almost certainly have been the case that NATO's nuclear forces and command and control infrastructure would have been degraded in the ways outlined above.

The physical task of terminating nuclear war not only included enforcing the ceasefire but also required the re-establishment of command and control and the return of nuclear weapons to storage or to *ante bellum* deployments.

The first requirement of nuclear war termination is that the appropriate high-level authority can be identified and that forces in the field should recognise it as such. Since in theatre nuclear war it would not be expected that the US political leadership would come under attack this need not have been a large problem, but for British and French forces it could have been. Even if the British political leadership survived it would not have had the means (that is the appropriate authentication codes) to impose high-level political control if the principal military command centres (where the codes were kept) had been destroyed.

The second requirement is that the message to cease hostilities goes out to all relevant field units. With degraded or destroyed command and control and command centres this would, if possible at all, have taken a considerable period, a period in which nuclear forces could continue to engage.

To actually impose a ceasefire of nuclear operations requires a highly diverse set of communication assets. Leahy's study of strategic nuclear war termination makes the following assessment:

> . . . if war termination is to be efficient, several kinds of communications will be needed: communications with forces to inform of the ceasefire arrangements and timing constraints; communications with attack warning and assessment sensors (to provide information on cease fire compliance, for example); communications with intelligence resources to inform the decision-making process leading to the initiation of war termination and the negotiations attendant on its implementation . . .; intragoverment communications to permit reaching a concensus on the necessity for and the acceptable terms of war termination; and intergovernmental communications . . . to initiate, negotiate and implement the termination.[103]

In the scenarios which have been discussed above many if not most of these communication systems in the European theatre would have been degraded or destroyed.

Even if communications could be re-established it would have been a major task to locate, identify, relock if necessary, and organise the return of perhaps thousands of dispersed nuclear weapons. NATO's nuclear weapons accounting system was probably not able to keep track of a nuclear arsenal which had been widely dispersed. Some of the weapons would have been used, others overrun, others destroyed or damaged, and still others lost or cut-off from authority during the conduct of war. Simply locating the weapons would have been hugely problematic, particularly if, as Bracken has argued, certain groups had chosen to keep nuclear weapons for their own political purposes[104].

The US Nuclear Capability Plan (NCP) and the Nuclear Weapon Accounting System (NWAS) were held at the four principal WWMCCS sites in Europe, and their British and French counterparts (held respectively at High Wycombe and Northwood in the UK and at Taverny and Val d'Oise in France) and would have been used as the basis for accounting the weapons. If any or all of these sites had been damaged or destroyed the task would have been greatly complicated.

The final task, that of relocking weapons which had been released required the matching of PAL codes to specific weapons. If Pirmasens (an above ground centre) had been destroyed finding the correct codes for the accounted weapons would have been greatly delayed. Taken together these arguments suggest that theatre nuclear war termination would have been a very difficult task.

CONCLUSIONS

Collective political decision-making in crisis and war was crucial to NATO's ability to fight as a cohesive whole and, if necessary, to order and control the use of nuclear weapons. Disagreements between the allies on any of the various aspects of authorising the use of nuclear weapons may have delayed or even forclosed NATO's collective nuclear operations.

The obstacles to collective decision-making on the use of nuclear weapons were formidable. The allies had to agree the need to transfer forces to NATO control, the desirability of initiating if necessary the use of nuclear weapons, the preparations for the use of nuclear weapons, the possibility of second (or nth) use, and the chosen weapons, nature, scale and timing of any use.

The political diversity of the NATO alliance and the evidence of past crises suggest that in practice NATO has not been able to reach timely collective agreement in crisis circumstances even when the use of force was relatively remote. Rather NATO's cumbersome peacetime decision-making bureaucracy has been shown to be an inadequate tool for crisis decision-making and the national interests of member nations have acted against consultation in the proper meaning of the term.

What emerges from an analysis of decision-making in crises is a clear hierarchy of nations within NATO led by the US and thereafter by the British, French and Germans. These four nations, and particularly the US, have been central to virtually all critical crisis decision-making in NATO in the post-war period. The likelihood appears to be that in crises NATO's multinational decision-making bodies (the NAC or DPC) would have become ineffective as decision-making authority shifted to the executives of NATO's leading member states.

Reticence amongst NATO members at any point in a crisis would have had not only had a profound effect upon collective decision-making but also upon the authorisation of the military measures deemed necessary to prepare for combat. The smooth transition to war (a process, it is worth recalling, NATO has never fully practiced), the chop, and preparations for the use of nuclear weapons would each have been greatly hampered by the lack of commitment or delayed commitment of NATO member nations. Decisions to cover the areas left undefended or underdefended by any lack of action on the part of individual or groups of nations would have needed to be quickly taken by those powers which remained committed to military measures.

It was probably at the point of authorising and deciding on the parameters of the use of nuclear weapons that differences between the NATO member nations would have been thrown into sharpest relief. In the final analysis the nuclear powers could have pushed on with a nuclear defence of NATO even if the other NATO members withdrew support, but at a high political cost.

Once committed to military action and the use of nuclear weapons NATO's political leaders would have been under pressure to allow military aims and logic to dictate the pace and shape of events. The contradiction between political and military goals in the nature and aims of using nuclear weapons could have created considerable tensions between the two. Political leader's lack of knowledge about military matters, the logic of NATO pre-planned mobilisation and preparations for the use of nuclear weapons, and the military desire for prompt and decisive force would have acted against political desires to limit the scale of nuclear use and political attempts to control or micromanage the conflict.

Military-political tensions might also have exposed differences between NATO and national decision-making creating considerable problems between national military forces trained and integrated into a NATO military force which may have been heading in one direction yet still responsive to national political leaders who may have wished to go in another. The autonomy of senior NATO commanders to raise the NATO alert status exemplified this dilemma.

If force – conventional or nuclear – had been used against the command and control infrastructure of NATO the gravity of all NATO's decisions with respect to nuclear forces would have been heightened still further at a time when much of NATO's apparatus for consultation and collective decision-making was being degraded or destroyed. National decision-making – primarily of the nuclear powers

– would have emerged as the dominant political process shaping NATO's nuclear actions at this point.

The military problems of interoperability, system limitations and vulnerability, and system performance in crisis and war provide further evidence of the difficulties NATO would have faced in seeking to implement the strategy of flexible response.

Careful analysis of the ability of NATO's command and control system to carry out the controlled and graduated use of nuclear weapons required by the strategy of flexible response – whilst unable without empirical evidence to reach any firm conclusions – does suggest that it would have been extremely difficult under all but the most benign of circumstances for NATO.

Conclusion

For more than 40 years NATO premised its defence upon the threat to use nuclear weapons. To make credible this threat and thus seek to ensure its deterrent efficacy NATO sought to develop a plausible strategy, the means to allow collective political control and collective military use of nuclear weapons, and a command and control infrastructure capable of supporting the nuclear weapons operations implicit in the strategy. This book has offered a critique of these efforts and its main conclusions are that NATO failed in these attempts and as a consequence took a terrible risk with the security of Europe. This is a bold claim – given that war within the European theatre was avoided – and requires elucidation.

The absence of war in Europe after 1945 can be explained by many factors of which the nuclear balance is but one. It has been no part of this study to attempt to determine the relative role of nuclear weapons in this process. My concern rather has been to seek to determine how well NATO has met the objectives she identified for herself in maintaining a deterrent based on a multinational nuclear force.

At the political level these objectives comprised tight political control of nuclear forces: the means to share nuclear decision-making on the development and deployment of nuclear weapons; and the means to provide central and collective political control of the use of nuclear weapons.

The foregoing has sought to identify the failures in these areas. Many of the nuclear forces operated by NATO were not subject to tight political control across the operational spectrum. Some systems such as American naval nuclear weapons and British nuclear weapons were subject to no physical control measures and high level political control was dependent upon no more than military obedience. Other systems – such as short range tactical nuclear weapons – were intended for use under conditions which precluded real-time political control.

With respect to political decision-making NATO developed an elaborate infrastructure for nuclear sharing which did indeed give political leaders of member nations input into some nuclear decisions. However, these formal peace-time arrangements obscured the reality that some states were more equal than others in terms of nuclear decision-making and that operational decision-making about the use of nuclear weapons remained both formally and practically in national rather than collective hands. The latter is an important point: in forty years NATO did not solve the question of how to achieve the collective control of nuclear forces.

In thinking through the issues faced by political leaders had the intention to use nuclear weapons arisen it became clear in this study, not least from evidence of past crises, that collective decision-making on issues of preparing to use and using nuclear weapons was a practical fiction. There were, quite simply, too many points of potential misperception, ambiguity, tension and disagreement to allow a collective decision in a 16 member organisation. Perhaps the most important insight was the evidence of powerful tensions between nuclear use and the cohesion of the alliance.

NATO's own arrangements recognised this in practice because it at no time agreed measures to seriously constrain the autonomy of the nuclear states within NATO, and was rebuffed or ran into practical difficulties when it attempted to do so.

At the military level NATO objectives included the articulation of a strategy which would underwrite the deterrent posture and provide the means to allow a multinational nuclear defence to be conducted. The trip-wire strategy of 1952 to 1967 committed Europe, East and West, to mutual anihilation in the event of Soviet aggression and was rejected in the 1960s as lacking credibility, and being inflexible and suicidal. Its replacement – flexible response – offered nuclear weapons as precise military instruments capable of meeting every criticism levelled at the trip-wire strategy. In the new strategy nuclear weapons could be chosen from a menu of short, medium and long range systems delivered at sea, from the ground and from the air. These weapons could be directed to specific objectives, both military, such as redressing conventional defeat, and political, such as demonstrating alliance resolve. Moreover, the strategy promised, and was premised on, the ability to offer these options at any point in a conflict, even after NATO had come under conventional or nuclear attack.

NATO met the force element of this strategy by deploying across the Western European theatre a wide range of weapons from small tactical systems for battlefield use [for example, nuclear artillery and mines] to Intercontinental Ballistic Missiles.

It provided command organisation through parallel national and NATO command structures. NATO readiness and operations were premised on the theatre-wide metamorphosis of the command structure from its peace-time to war-time form. This process involving sixteen nations, hundreds of command centres, thousands of conventional and nuclear weapons, and millions of personnel was never practiced. The first test of NATO's ability to undergo this profound transformation would have come at the moment of a serious crisis or conflict with the Soviet Union.

NATO met the command and control demands of the strategy by deploying the infrastructure that has been described at length in this study. It has been the focus of this analysis to argue that NATO's command and control infrastructure would not have enabled NATO to conduct the operations required by the strategy of flexible response. Problems derived from lack of overall systems architecture, lack of interoperability, system limitations and vulnerabilities, and system performance in crisis and war suggest that flexible response was simply too demanding and too elaborate.

These collectively are the grounds for arguing that NATO failed to meet its own objectives and as a consequence took a great risk with European security. A risk because my analysis and that of others – including Ball, Bracken, Charles, Kelleher and Steinbruner – suggests that NATO risked paralysis, chaos, loss of political control over the use of nuclear weapons and rapid nuclear escalation in the event of a serious crisis or conflict. Bracken's suggestion that it was precisely this predisposition to chaos and loss of control which gave NATO's deterrent credibility implies that NATO knowingly constructed a system designed to fail catastrophically or at least knowingly allowed such a system to evolve. If true NATO would be guilty of the gravest and most irresponsible gamble.

It is more convincing to my mind to argue that between 1967 and 1990 NATO allowed a huge gap to open between the operations implicit in her nuclear strategy and her ability to conduct those operations. The reasons for this are quite clear.

At the political level peacetime concerns about equity within the alliance, the need for clear mechanisms for the collective control of nuclear weapons, and the need for observable alliance cohesion on the nuclear question subsumed crisis and wartime realities and led to the development of elaborate fora and procedures to meet primarily peacetime needs. In crises the evidence is that collective fora are inappropriate bodies for necessary decision-making, that collective procedures and mechanisms cannot solve differences of national interest, perception, analysis and interpretation between members states, and that decision-making shifts naturally and rapidly to national executives. The decision to use nuclear weapons cannot, at least given present constitutional arrangments in the nuclear member states, be shared.

At the level of strategy flexible response was shaped by think-tanks to meet criticisms of the trip-wire strategy and the demands of abstract deterrence calculations. It was accepted as an idea because it met peacetime political concerns and was introduced without the supportive infrastructure to make it feasible being in place. One cannot observe in retrospect in the period prior to or even after 1967 new systems being introduced to provide this capability. Flexible response was introduced into a context of systems designed for trip-wire top-down general release of nuclear weapons, and NATO did not subsequently devise or deploy the means to fully support the strategy.

At the military level the command structure evolved as a means of attempting to unify sixteen national armed forces spread across a vast geographic region of multilingual states of different political preferences. Parallel command structures were needed to allow armed forces to meet national as well as NATO policies. Since NATO did not go to war, while states did, national policies generally took preference. The plans for transformation from national to NATO posture became more elaborate and more complex because they were not tested. They were not tested as a whole because of cost, domestic opposition within member states, the risk of public difficulties or failure if exercised, and because the process risked being provocative and escalatory.

NATO's command and control system lacked an overall systems architecture and lacked theatre or MNC interoperability in a range of mission areas because of national parochialism which insisted that national systems be allowed to proliferate and because of financial constraints which led to system cancellations, delays and specification compromises.

Criticism of this kind is however useful only in so far as there are lessons to be learned from it. The next question is thus whether there are lessons for the present greatly changed context of European security from NATO's experience of operating nuclear weapons between 1952 and 1990. The reasons for arguing that lessons of the past are still relevant lie in arguments about the continuity of threats to NATO security in the international system and in the questions of whether NATO will remain in existence and whether it will continue to function as a multinational nuclear alliance.

To deal first with the issue of threat. The end of the Cold War has undoubtedly transformed the European context but it has not brought an end to threats to NATO member states. Analysts have identified at least four threats which NATO may need to face. Firstly the issue of latent Russian power. Russia remains a large military state in possession of a large nuclear arsenal which is still one of the primary concerns of analysts and governments in NATO member states. Whether Russia remains politically stable, the complexion of its leadership, and the nature of Russia's relations with the west will all be key determinants of concerns about the risks of latent Russia military and particularly nuclear power.[1]

A second threat is associated with new fault-lines which some analysts argue are opening in relations between the west and other states or regions of the world. Such concerns have been explored with varying degrees of conviction by scholars such as Huntington who points to a 'clash of civilisations' which may shape the future security context.[2]

A third concern lies in the divisions between the prosperous developed, industrialised and technically advanced 'North' and the less prosperous, underdeveloped, less industrialised and less technically advanced 'South'. In this analysis the North and South, or if you prefer the west and the Third World, may clash over issues as diverse as population migration, access to natural resources, the spread of weapons of mass destruction [biological, chemical and nuclear], and political and social instability.[3]

Finally, the west has also begun to articulate concerns about an 'Islamic threat' with particular resonance in Europe given the close proximity of the Middle East and North Africa and the presence of large Islamic populations within many European states.[4]

Whatever the respective force of these arguments it is difficult to argue that NATO lives in a threat-free context, despite the end of the Cold War.

NATO will certainly continue to exist in the short-to-medium term albeit perhaps in a changed form given the possible accession of at least some former-Warsaw Pact states.[5] The last question – whether it will remain a multinational nuclear alliance – can also be answered positively. Western analysts and governments have set out many arguments to justify and explain the continued reliance on nuclear weapons. These include the arguments that nuclear weapons exist and cannot be disinvented; that even if complete nuclear disarmament could be agreed today it would still take many years to implement during which time nuclear weapons would remain a feature of international relations; the difficulties of agreeing complete nuclear disarmament; the role of nuclear weapons in the regulation of international conflict and their usefulness in times of uncertainty; the risks of the proliferation of weapons of mass destruction; and, the risk of political regression in Russia, China and elsewhere.[6] No-one is seriously arguing at this moment that the United States, United Kingdom or France will eliminate their nuclear weapons.

Given that NATO will remain a multinational nuclear alliance for the forseeable future in a changed but threat-potential international context the lessons drawn above from the 1952–90 period can offer useful insights.

In the post-Cold War context the NATO nuclear command and control environment has become comparatively benign. In the absence of a continental threat of invasion the military dynamics explored above have been changed and the time urgency dimension of nuclear operations have been largely eliminated. The removal of short-range nuclear weapons have also removed the key element in escalatory pressures due to the dispersal and the loss of political control of nuclear weapons. The shift to a reliance on longer-range nuclear weapons [missile and air-borne] has greatly improved the political contollability of NATO's nuclear weapons. Finally, the end of the flexible response strategy has relieved NATO of an unconvincing, over-elaborate and dangerous strategy. All of these events are to be greatly welcomed.

However, there are a number of outstanding issues which NATO, and particularly the European members of NATO, ought to address irrespective of whether the future sees a return to a continental threat or not. A study such as this, which has focussed exclusively on the past, cannot offer ready-made answers for the present, nor does it presume to do so. It can however point out focal areas for alliance and member state attention the better to enable NATO to face an uncertain future, benign or otherwise.

On the assumption that NATO will remain a multinational nuclear alliance – by which I mean that nuclear weapons will continue to play a role in NATO strategy and operations – there are at least four principal areas of concern for NATO member states:

NUCLEAR SHARING

NATO has not solved the problem of how to ensure that nuclear weapons are used only with the assent of member states. At present and in the past arrangements allowed the nuclear powers to use nuclear weapons on behalf of the alliance with or without their consent. Member states must address the question of whether this remains a desirable state of affairs, if indeed it ever was. Given that the possibility of the alliance using nuclear weapons has receded for the present to the dim and distant horizon this is perhaps not a question which will attract much attention, not least because of its complexity, sensitivity and divisiveness. However, this is an issue which NATO needs to address not least because NATO may find itself in a context in which weapons of mass destruction have proliferated or in which a continental threat has re-emerged. The evidence suggests that deepening and extending alliance participation in nuclear decision-making or rethinking issues of negative or positive control – that is giving all member states the ability to veto use or requiring complete agreement to authorise use – are either side issues or impractical given the consitutional realities in the nuclear member states. Non-nuclear alliance members must it seems either live with this situation or question whether nuclear weapons any longer have a place in the alliance given their inability to control their use. The nuclear states face similar questions at least to the extent that they cannot control the actions of the other nuclear member states. The options it would appear are thus: accept the status quo; rethink the nuclear presence in the alliance; or pressure the nuclear states to accept forms of collective control – at least in the alliance context – which the nuclear states have hitherto always rejected. It is on the basis of resolving these issues that appropriate fora and mechanisms for the political control of nuclear weapons in the alliance of the future could then be decided.

STRATEGY

NATO's new strategy must close the gap between nuclear weapons operations implicit in the strategy and NATO's ability to carry out the operations implied. This principle is as sound for operations within the theatre as it is for those outside it. Important steps along this path are the need to limit the role of nuclear weapons by recognising them solely as weapons of last resort, separating them from conventional armed forces and conventional battle plans, and deploying the necessary command and control infrastructure to support their operations.

COMMAND AND CONTROL INFRASTRUCTURE

This study has identified widespread and deep-seated problems with the nature and performance of NATO's nuclear command and control infrastructure. It is important for NATO to address these issues. NATO thinking should be shaped by three ideas: tight political control of nuclear weapons; sustainable connectivity; and counter-escalation. The key areas of concern are firstly the need to provide and co-operate on theatre and MNC systems architecture in different mission areas to address the lack of interoperability.[7] This will require co-operation on research and development, the procurement of systems, and the will to invest resources in this relatively low profile area of military expenditure. Secondly the need to enhance the safety and security of nuclear weapons by the retrospective introduction of physical control measures on all nuclear weapons systems which do not presently have them and the improvement and sharing between nuclear states of present technology. In the present context all of the arguments against fitting nuclear weapons with physical security devices can be discounted, if indeed they were ever fully convincing.

Thirdly, investment should be made in sustainable connectivity between nuclear forces and political authority. This has been made far more straightforward by the reduction of types and numbers of weapons deployed by the alliance and by the shift to reliance on longer-range nuclear weapons.

Finally, command and control predispositions and vulnerabilities which make the use of nuclear weapons more likely and contribute to the dynamics of use should be replaced by predispositions and system robustness which act in a de-escalatory manner. In other words no feature of the nuclear command and control infrastructure or the weapons it controls should make NATO states more likely rather than less likely to use nuclear weapons. The lessons to be learnt here draw on the escalatory mechanisms and the difficulties of terminating use which were discussed above. Counter-escalatory ideas include: better and outside-combat-theatre weapons storage; improved connectivity robustness; improved physical control measures; remote relocking of released weapons; improved weapons accounting; and separation of conventional and nuclear systems. For as long as NATO remains a nuclear alliance its nuclear weapons should be as safe, secure and tightly controlled politically as is possible.

A EUROPEAN SECURITY IDENTITY

The three issues outlined immediately above are also relevant to any notions European member nations have about a distinctly European security identity be it within the European Union, the WEU or some other arrangement. A nuclear presence in such an arrangement faces the nuclear sharing issues outlined, while ideas of joint military operations face the command and control issues also outlined. A third problem relates simply to the absence of European command and control infrastructure in critical mission areas which this study has detailed. The assessment that without the United States European members lack intelligence gathering and analysis capabilities [strategic and tactical], attack warning and attack assessment capabilities and have shortfalls in most other command and control mission areas [with the arguable exception of air defence] not least on issues of connectivity and interoperability, together with the foregoing, puts into proper context any notions of a European security identity and a European nuclear deterrent force.

Command and control emerged as an important branch of strategic analysis in the 1980s. Although the security context has changed there is still a great deal to be learned from understanding the associated issues. It remains to be seen whether NATO member states will give appropriate attention to these concerns.

Notes and References

Introduction

1. For the reasons behind the development of the strategy of flexible response and the way in which American and European consensus on the new strategy was achieved see David N. Schwartz, *NATO's Nuclear Dilemmas* (Washington DC: Brookings Institution, 1983) 136–92.
2. NATO Information Service, *The North Atlantic Treaty Organisation: Facts and Figures* (Brussels: NATO Information Service, 1989) 139.
3. Details about the nuclear weapons stockpiles of the USA, UK and France are taken from William Arkin and Richard Fieldhouse, *Nuclear Battlefields* (Cambridge, Mass.: Ballinger Press, 1985) 37–63; and Catherine Kelleher, 'NATO Nuclear Operations' in Ashton Carter, John Steinbruner and Charles Zracket (eds) *Managing Nuclear Operations* (Washington DC: Brookings Institution, 1987) 448.
4. Thomas Cochran, William Arkin and Milton Hoenig, *Nuclear Weapons Databook Vol 1: US Nuclear Forces and Capabilities* (Cambridge, Mass.: Ballinger Press, 1984) 42–54.
5. Ibid., 136–7, 306–7.
6. See Greg Todd 'C_1 Catharsis', *Defense Analysis*, vol. 4, no. 3 (September 1988) 203.
7. Martin Van Creveld, *Command in War* (Cambridge, Mass.: Harvard University Press, 1985) 1.
8. William Arkin and Richard Fieldhouse 'Nuclear Weapon Command, Control and Communications' in *SIPRI Yearbook 1984* (London and Philadelphia: Taylor and Francis, 1984) pp. 455–7.
9. Bruce Blair, *Strategic Command and Control* (Washington DC: Brookings Institution, 1985) 1.
10. This definition is derived from that of Paul Bracken in *The Command and Control of Nuclear Forces* (New Haven, Conn.: Yale University Press, 1983) 3. The definition has been expanded to make explicit the function of decision-making and both national command authorities and military commanders have been specified to make clear the involvement of both political and military decision-makers in nuclear command and control.
11. Herbert D. Bennington, 'Command and Control for Selective Response', in Klaus Knorr and Thornton Read (eds) *Limited Strategic War* (New York: Praeger, 1962) 119.
12. William Arkin and Richard Fieldhouse, op. cit., 456.
13. Bruce Blair, op. cit., 4.
14. Herbert D. Bennington, op. cit., 119.
15. J.M. Legge, *Theatre Nuclear Forces and the NATO Strategy of Flexible Response* (Santa Monica, California: RAND Corporation, April 1983) R-2964-FF, 40–1.

16. Ashton Carter, John Steinbruner and Charles Zracket, *Managing Nuclear Operations* (Washington DC: Brookings Institution, 1987) 7.
17. The best is Milton Leitenberg's chapter entitled 'Background Materials on Tactical Nuclear Weapons (Primarily in the European Context)', in Frank Barnaby (ed.) *Tactical Nuclear Weapons European Persectives*, SIPRI, (London: Taylor and Francis, 1978) 3–136. Although not a specific study of command and control the piece contains a great deal of interesting material.
18. Desmond Ball, *Can Nuclear War be Controlled?* (IISS, 1981) (London: Adelphi Paper 169).
19. Bruce Blair, op. cit.
20. Paul Bracken, op. cit.
21. John Steinbruner, 'National Security and the Concept of Strategic Stability', *Journal of Conflict Resolution*, vol. 22, 3 (1978) 411–28; and John Steinbruner, 'Nuclear Decapitation', *Foreign Policy* (Winter 1981–2) 16–28.
22. M. Blumenson and J. L. Stokesbury, *Masters of the Art of Command* (Boston, Mass.: Houghton Mifflin, 1975).
23. Robert Coulam and Gregory Fischer, 'Problems of Command and Control in a Major European War', in Robert Coulam and Richard Small (eds) *Advances in Information Processing in Organisations Vol 2* (New York: JAI Press Inc, 1985) 211–73.
24. John Cushman, *Command and Control of Theatre Forces: Adequacy* (Cambridge, Mass: Harvard University Press, 1983).
25. Paul Stares, *Command Performance: The Neglected Dimension of European Security* (Washington DC: Brookings Institution, 1991).
26. C. Kenneth Allard, *Command, Control and the Common Defense* (New Haven, Conn.: Yale University Press, 1990).
27. Daniel Charles, *Nuclear Planning in NATO* (Cambridge, Mass.: Ballinger Press, 1985).
28. Catherine Kelleher, op. cit., 445–69.
29. Desmond Ball, *Controlling Theatre Nuclear War* (Canberra, Australia: Australian National University, 1987) (Working Paper 127).
30. Paul Bracken, op. cit., 130.
31. Ibid.
32. Daniel Charles, op. cit., 151–2.
33. Ibid., 6.
34. Catherine Kelleher, op. cit., 468.
35. Ibid.
36. Ibid.
37. Desmond Ball (1987), op. cit., 3.
38. Ibid., 36.
39. Ibid., 14.
40. Crisis in the context of this book refers to an adversarial international security crisis defined as a period of change or decision in the relationship between adversarial nations or alliances which threatens to tranform the nature of that relationship from peace to war. See Coral Bell, *The Conventions of Crisis* (Oxford: Oxford University Press, 1971) 6–9.

41. Ashton Carter, John Steinbruner and Charles Zracket (eds), op. cit., 9.
42. Amongst many others see Graham Allison, *The Essence of Decision: Explaining the Cuban Missile Crisis* (Boston, Mass.: Little, Brown, 1971); Michael Brecher, 'Towards a Theory of International Crisis Behaviour', *International Studies Quarterly*, vol. 21, 1 (1977) 39–74; Ole Holsti, 'Theories of Crisis Decision-Making', in Paul Lauren (ed.) *Diplomacy* (Free Press, 1979) 99–136; and, Richard Ned Lebow, *Between Peace and War* (Baltimore, MD.: Johns Hopkins University Press, 1981).
43. Desmond Ball (1981 and 1987), Bruce Blair (1985), and Paul Bracken (1983), for example, all use this approach.
44. Paul Bracken, 'The Political Command and Control of Nuclear Forces', *Defense Analysis*, vol. 2, 1 (1986) 11.
45. Robert Coulam and Gregory Fischer, op. cit., 213–4.
46. G.D. Foster, 'Contemporary Command and Control Theory and Research: The Failed Quest for a Philosophy of Command', *Defense Analysis*, vol. 4, 3 (September 1988) 201–28.
47. Ashton Carter, John Steinbruner and Charles Zracket (eds), op. cit., 3.
48. See Jeffrey T. Richelson, *The US Intelligence Community* (Cambridge, Mass.: Ballinger Press, 1985; and, Richard K. Betts, *Surprise Attack* (Washington DC: Brookings Institution, 1982) especially 153–254.
49. For an explanation of force coupling see Paul Bracken (1983), op. cit., 54–9.

1 Political Control of Nuclear Weapons In NATO

1. *Authority to Order the Use of Nuclear Weapons* (Washington DC: Congressional Research Service, House of Representatives, Subcommittee on International Security and Scientific Affairs of the Committee on International Relations, 1 December 1975) 1.
2. Theatre Nuclear Forces. Throughout the book the term theatre nuclear forces or the acronym TNF is used to include tactical and theatre nuclear forces, unless otherwide explicitly stated.
3. *The Atlantic Alliance* (Washington DC: Hearings, The Committee on Government Operations, US Senate, 89th Congress, Part 2, May 1966) 69–86.
4. Lawrence Freedman, *The Evolution of Nuclear Strategy* (London and Basingstoke: Macmillan Press, 1982) 69.
5. Ibid., 68.
6. Milton Leitenberg, 'Background information on tactical nuclear weapons', in *Tactical Nuclear Weapons: European Perspectives*, SIPRI, (London: Taylor and Francis, 1978) 12.
7. *The Atlantic Alliance*, op. cit., 69.
8. Because of the lack of strategic range delivery systems at the time strategic nuclear weapons had been forward based in some European countries, notably the UK, since 1948. See Robert Jackson, *Strike Force: The USAF in Britain Since 1948* (Bury St Edmonds, Suffolk: Robson Books, 1986).

9. William Arkin and Richard Fieldhouse, *Nuclear Weapons Databook Vol I* (Cambridge, Mass.: Ballinger Press, 1984); and Leitenberg (1982), op. cit., 109–16.
10. David Schwartz, *NATO's Nuclear Dilemmas* (Washington DC: Brookings Institution, 1983) 263.
11. Robert E. Osgood, *NATO: The Entangling Alliance* (Chicago, Ill.: University of Chicago Press, 1962) 109.
12. J. M. Legge, op. cit., (April 1983) 6.
13. Ibid.
14. Ibid., 7.
15. Simon Duke, *US Defence Bases in the UK* (London: Macmillan Press, 1987) 31–40.
16. Carl H. Amme, *NATO Without France* (Stanford, California: Hoover Institution, 1967) 153–72. See also Guy De Carmoy, *The Foreign Policies of France 1944–68* (Chicago, Ill.: University of Chicago Press, 1970) 283–92.
17. *NATO Final Communiqués 1949–74* (Brussels: NATO Information Service, 1975) 113.
18. Milton Leitenberg, op. cit., 15.
19. Peter Stein and Peter Feaver, *Assuring Control of Nuclear Weapons* (Boston, Mass.: University Press of America, 1987) CSIA Occasional Paper Series, 27.
20. Irving Heymont, 'The NATO Bilateral Forces', *Orbis*, vol. 9, 4 (1966) 1031–3.
21. Milton Leitenberg, op. cit., 15.
22. Irving Heymont, op. cit., 1033.
23. Robert E. Osgood, *Nuclear Control in NATO* (Washington DC: Washington Center for Foreign Policy Research, 1962) 21.
24. These concerns have been explored more fully in Klaus Knorr, 'Nuclear Weapons: "Haves" and "Have-Nots"', *Foreign Affairs*, vol. 36, 1 (April 1957) 167–78 and Albert Wohlstetter, 'Nuclear Sharing: NATO and the N+1 Country', *Foreign Affairs*, vol. 39, 3 (April 1961) 355–87.
25. Milton Leitenberg, op. cit., 15.
26. P.M. Gallois, 'New Teeth for NATO', *Foreign Affairs*, vol. 39, 4 (October 1960) 67–80.
27. Robert Osgood, *NATO: The Entangling Alliance*, op. cit., 284.
28. To follow this up see ibid., 284–5.
29. Ben Moore, *NATO and the Future of Europe* (New York: Harper and Row, 1958); see also Glen H. Snyder, *Deterrence and Defense* (Princeton, N.J.: Princeton University Press, 1961) 174–92.
30. This idea is explored in Alistair Buchan, *NATO in the 1960s* (London: Weidenfeld & Nicolson, 1960) see also Snyder, op. cit., 174–92.
31. Robert Osgood, *NATO: The Entangling Alliance*, op. cit., 295.
32. See ibid., 295–307 for a more detailed discussion. See also Alistair Buchan, 'The Reform of NATO', *Foreign Affairs*, vol. 40, 2 (January 1962).
33. David Schwartz, op. cit., 82. Also whole chapter, 82–135; Wilfried Wohl, 'Nuclear Strategy in NATO and the MLF', *Political Science Quarterly*, vol. LXXX, 1 (March 1965) 89–109; John Slessor, 'The Case for a

Multinational Nuclear Strike Force', in K. H. Cerny and H. W. Briefs (eds) *NATO in Quest of Cohesion* (Stanford, California: Hoover Institution, 1965) 239–56.
34. Francis Beer, *Integration and Disintegration in NATO* (Cincinatti: Ohio State University Press, 1969) 67–72.
35. Quoted in David Schwartz, op. cit., 134.
36. The text of the North Atlantic Treaty is reproduced in *NATO Facts and Figures* (Brussels: NATO Information Service, 1984) 264–6.
37. Harland Cleveland, *NATO: The Transatlantic Bargain* (New York: Harper & Row, 1970) 14. See particularly Chapter 2, 'The Golden Rule of Consultation', 13–33.
38. The 1951 Committee members were W. Averell Harriman (USA), Jean Monnet (France) and Edwin Plowden (UK).
39. The quotation from the 1951 NATO Committee on the North Atlantic Council is quoted in the text of the 1956 Report of the Committee of Three on Non-Military Co-operation in NATO, which is reproduced in *NATO Facts and Figures*, 1984, op. cit., 275–6.
40. The three were Harvard Lange, Gaetano Martino and Lester B. Pearson, respectively the Foreign Ministers of Norway, Italy and Canada.
41. Report of the Committee of Three on Non-Military Co-operation in NATO, which is reproduced in *NATO Facts and Figures*, 1984, op. cit., 275.
42. Ibid., 1984, 277.
43. See Henry Kissinger, *Years of Upheaval* (London: Weidenfeld & Nicolson, 1982) 192–3. The backgound to the 'Year of Europe' and the deterioration of relations between America and Europe during this period is described by Kissinger in Chapter 5, Ibid., 128–95. See also Alfred Grosser, *The Western Alliance: European-American Relations Since 1945* (London: Macmillan Press, 1980) Chapter 10, 263–83.
44. The Declaration on Atlantic Relations is reproduced in *NATO Facts and Figures*, op. cit., 292–4. See paragraph 11.
45. Roy W. Stafford, 'Defence Planning in NATO: A Consensual Decision-Making Process', in R. L. Pfaltzgraff and U. Ra'anan (eds) *National Security Policy: The Decision-Making Process* (Hamden, Conn.: Archon Books, 1984) 153–4.
46. Richard Neff, 'NATO Political Consultation: Fact or Myth?', *NATO Review*, vol. 23, 1, (January 1975) 9.
47. Frederic L. Kirgus, 'NATO Consultations as a Component of National Decision-Making', *American Journal of International Law*, vol. 73, 3 (July 1979) 404.
48. Ibid., 405.
49. Ibid.
50. Joachim Jaenicke, 'Political Consultation – Reality or Formality', *NATO Letter* (April 1969) 16.
51. J. M. Legge, op. cit., 14. The exact text of the Athens Guidelines remains classified.
52. *NATO Communiqués 1949–74* op. cit., 144.
53. Ibid., 151.

54. 'The Nerve Centre of United States Nuclear Strategy', *NATO Letter* (May 1966) 19.
55. John C. Burney, 'Nuclear Sharing in NATO', *Military Review* (June 1969) 63.
56. Harland Cleveland, op. cit., 53–4.
57. David Schwartz, op. cit., 183–5.
58. *NATO Communiqués 1949–74* op. cit., 180.
59. Paul Buteux, *The Politics of Nuclear Consultation in NATO 1965–80* (Cambridge: Cambridge University Press, 1983) 57–8. Buteux's book contains a detailed consideration of nuclear consultations in NATO. On the Nuclear Planning Group in particular see 39–213.
60. *US Security Issues in Europe: Burden Sharing and Offset, MFBR and Nuclear Weapons* (Washington DC: Staff Report, Committee on Foreign Relations, US Senate, 93rd Congress, 2 December 1973) 19–20.
61. Ibid.
62. Roy W. Stafford, op. cit., 154–5.
63. Catherine Kelleher, 'NATO Nuclear Operations', in Ashton Carter, John Steinbruner, and Charles Zracket, (eds) *Managing Nuclear Operations* (Washington DC: Brookings Institution, 1987) 464.
64. Milton Leitenberg, op. cit., 15.
65. *NATO Communiqués 1949–74*, op. cit., 144.
66. John Schlesinger, *Theatre Nuclear Force Posture in Europe* A Report to US Congress in Compliance with Public Law 93-365 (Washington DC: US Congress, 1974) 25.
67. *NATO Communiqués 1949–74*, op. cit., 186.
68. J.M. Legge, op. cit., 17.
69. *NATO Communiqués 1949–74*, op. cit., 203.
70. Initial use was intended to mean the first use of nuclear weapons by NATO either before or in response to any use of nuclear weapons by the Warsaw Pact.
71. J.M. Legge, op. cit., 18.
72. Ibid.
73. Ibid, 19.
74. *NATO Communiqués 1949–74*, op. cit., 228.
75. J.M. Legge, op. cit., 26.
76. Ibid., 27.
77. Ibid.
78. Ibid.
79. *NATO Communiqués 1975–80* (Brussels: NATO Information Services, 1981) 33.
80. *NATO Communiqués 1949–74*, op. cit., 242.
81. Daniel Charles, *Nuclear Planning in NATO* (Cambridge, Mass.: Ballinger Press, 1987) note 2, 130.
82. *NATO Facts and Figures* (Brussels: NATO Information Service, 1989) 224–5.
83. Author's interviews.
84. Author's interviews.
85. *NATO Facts and Figures*, op. cit., 204.

86. John Anderson, 'The Evolution of NICS', *Signal* (April 1979) 15–6.
87. John Guthrie, 'Underground Command Posts', *International Combat Arms* (May 1987) 19.
88. Harold A. Kissinger, 'Organisation for Crisis Management in NATO', *Signal* (March 1975) 68.
89. Ibid., 69.
90. *NATO Facts and Figures*, 1984, op. cit., 204.
91. Harold Kissinger, op. cit., 67.
92. *NATO Facts and Figures*, 1984, op. cit., 204.
93. Peter Hill-Norton, 'Crisis Management', *NATO Review* (October 1976) 8.
94. Ibid.
95. John Anderson, op. cit., 16.
96. The ACE-HIGH communications network is the subject of more detailed consideration in Chapter 3.
97. John Anderson, 'NATO's Communications Requirement and the Role of NICSMA', *Journal of the Royal Signals Institute*, vol. 15: 4 (Spring 1982) 159.
98. Peter Jenner, 'NATO-wide Communications Improved', *NATO Letter* (April 1970) 7–8.
99. John Anderson (1982), op. cit., 160.
100. R. N. B. Freeman, 'The NATO Integrated Communications System – Stage I', *Journal of the Royal Signals Institute*, vol. 17, 3 (Winter 1985) 156.
101. *NATO Facts and Figures*, 1984, op. cit., 180.
102. J. A. D. Francis, 'The NATO Integrated Communications System', *Journal of the Royal Signals Institute* vol. 13, 5 (Summer 1978) 17.
103. John Anderson (1979), op. cit., 16.
104. R. N. B. Freeman, op. cit., 156.
105. John Anderson (1979), op. cit., 17.
106. R. N. B. Freeman, op. cit., 156.
107. 'NATO ISVN', *Jane's Military Communications* (New York: Jane's Publications, 1988) 720.
108. Ibid., 719.
109. G. D. Hingorami and R. Brand, 'Architectural Framework for the Evolution of the NATO Integrated Communications Systems', *Signal* (October 1985) 56.
110. CIP-67 is the acronym for the NATO Communications Improvement Programme which began in 1967. It provides a line-of-sight radio network in the NATO central region. The system of 60 nodes became operational in 1983.
111. G. D. Hingorami and R. Brand, op. cit., 56.
112. *NATO Facts and Figures*, 1984, op. cit., 180.
113. G. D. Hingorami and R. Brand, op. cit., 56.
114. These systems are discussed in Chapter 2.
115. David Williams, 'NATO Merges Planners, Procurers to End Communications Patchwork', *Defense News* (19 May 1986) 4.
116. *Jane's Military Communications 1988*, op. cit., 720.
117. G. D. Hingorami and R. Brand, op. cit., 60.

118. *NATO Facts and Figures* (Brussels: NATO Information Service, 1989) 288–9.

2 Command and Control in NATO

1. *NATO Facts and Figures* (Brussels: NATO Information Service, 1989) 337–40.
2. Ibid., 339.
3. Frank A. Partlow, 'The NATO Military Committee and the International Military Staff: Some Rationales and a Proposal for Reorganisation', *RUSI Journal*, vol. 119 (September 1974) 31–2.
4. Roy W. Stafford, 'Defence Planning in NATO: A Consensual Decision-Making Process', in R. L. Pfaltzgraff and Uri Ra'anan (eds) *National Security Policy: The Decision-Making Process* (Hamden, Conn.: Archon Books, 1984) 156.
5. Elliot Vandervanter, *Some Fundamentals of NATO Organisation* (Santa Monica, California: Rand Corporation, 1963) 47, RAND Memorandum RM-3559.
6. 'The Transfer of SHAPE to Belgium', *NATO Letter* (December 1966) 12–3.
7. William Arkin and Richard Fieldhouse,op. cit., 1985, 219, 227–8.
8. The information about the command structure of ACE was taken from *Command Structure of ACE* (Brussels: ACE Backgrounder Briefing) SHAPE Headquarters, B31-0787, 10 July 1987.
9. Terry Gander, *Encyclopaedia of the Modern British Army* (Cambridge, UK: Patrick Stephens Limited Press, 1983) 17.
10. William Arkin and Richard Fieldhouse, op. cit., 236–45.
11. Franz-Jospeh Schulze, 'Command, Control and Communications in NATO', in W. Kaltefleiter and Ulrike Schumacher (eds) *Conflicts, Options, Strategies in a Threatening World* (Kiel: Institute of Political Science, Christian Albrechts University, 1981) 214.
12. *NATO Facts and Figures*, 1989, op. cit., 346–8.
13. The exceptions are the Airborne Early Warning Aircraft which are permanently assigned to NATO command.
14. D. Wood, 'Control of the Eastern Seaways', *Interavia*, 1 (1980) 44.
15. Even in the case of France there are exceptions, see Chapter Five.
16. 'SHAPE Communications', *Signal* (December 1959) 7.
17. Anne Sington, 'French Link in a NATO Communications System', *NATO Letter* (April 1965) 11.
18. 'SHAPE Communications', op. cit., 8.
19. Kenneth F. Zitzman, 'International System Problems as Exemplified by NATO Project ACE-HIGH', *Signal* (November 1960) 45.
20. Anne Sington, op. cit., 11.
21. 'ACE-HIGH Transmission System', *Jane's Military Communications 1988* (New York: Jane's Publishing Company, 1988) 717.
22. G. D. Hingorani, 'Digital Replacement of the ACE-HIGH Transmission System', *Signal* (December 1980) 61–4.
23. Anne Sington, 'NATO Satcom', *NATO Letter* (March 1968) 16.

24. 'NATO HF Communications Network' (New York) *Jane's Military Communications*, 1988, op. cit., 718.
25. G.D. Hingorani and R. Brand, 'Architectural Framework for the Evolution of NATO Integrated Communications System', *Signal*, October 1985, 56. See also Paul Stares, *Command Performance* (Washington, D.C.: Brookings Institution, 1991) 164.
26. 'NATO's Shelterised HF Communications System', *Jane's Defence Review*, vol. 3, 1 (1982) 85.
27. Author's correspondence with SHAPE headquarters, 22 November 1989.
28. Dennis C. Marquis, 'New Directions in NATO C^3', *Signal* (December 1982) 39–42.
29. J. Sochaczewski, 'NATO Command and Control and Information Systems', *Military Technology* (October 1988) 31–7.
30. James Schultz, 'Northern Europe to Receive New Battle Management System', *Defense Electronics* (May 1985) 64–8.
31. Ibid., 33.
32. P. Harding, 'C^3I Supporting the Commanders Game Plan', *Signal* (October 1987) 30.
33. 'NATO Looks Towards 2008', *Defence Electronics and Computing* (supplement to *International Defence Review*), 12 (1989) 152.
34. Jon L. Boyes, 'BICES: Linking NATO Intelligence', *Signal* (October 1987) 19–20.
35. Stephen Broadbent, 'Joint-STARS: Force Multiplier for Europe', *Jane's Defence Weekly* (18 April 1987) 729–33.
36. Wavell acts both as an information and intelligence system. See Rupert Pengelley, 'HEROS and Wavell: Battlefield ADP Enters a NEW Era', *International Defence Review*, 10 (1986) 1459–64.
37. 'NATO Looks Towards 2008', op. cit., 152.
38. A.L. Meier, 'BICES: A Central Region Perspective', *International Defence Review*, 10, (1986) 1445–9.
39. *NATO Facts and Figures*, 1989, op. cit., 275.
40. 'NATO Set to Upgrade SHAPE System', *Jane's Defence Weekly* (October 1987) 16.
41. 'New NATO C^3 System', *Defence* (August 1987) 7.
42. 'NATO Set to Upgrade SHAPE System', op. cit.
43. 'NATO Looks Towards 2008', op. cit., 151–3.
44. G.D. Hingorani and R. Brand, op. cit., 56.
45. 'The Nerve Centre of US Nuclear Strategy', *NATO Letter* (May 1966) 19–20.
46. Desmond Ball, 'The Development of the SIOP', in Desmond Ball and Jeffrey Richelson (eds) *Strategic Nuclear Targeting* (Ithaca, New York: Cornell University Press, 1986) 57–83.
47. Ibid.
48. J.M. Legge, op. cit., April 1983, 14.
49. Lawrence Freedman, 'British Nuclear Targeting', in Ball and Richelson, op. cit., 215.
50. William Arkin and Richard Fieldhouse, op. cit., 96.
51. Lawrence Freedman, 'British Nuclear Targeting', *Defense Analysis*, vol. 1, 2 (1985) 81–99.

52. Desmond Ball, *Targeting for Strategic Deterrence*, Adelphi Paper No 185, IISS, (Summer 1983) 16.
53. Catherine Kelleher, 'NATO Nuclear Operations', in A. Carter, J. Steinbruner and C. Zracket (eds) *Managing Nuclear Operations* (Washington DC: Brookings Institution, 1987) 450.
54. In November 1991 the last US Poseidon SSBN assigned to NATO and based at Holy Loch in Scotland was withdrawn. See 'USN exits Holy Loch', *Jane's Defence Weekly* (23 November 1991) 987. Its role has been assumed by Trident.
55. J. M. Legge, op. cit., 20.
56. *Conventional – Nuclear Operations*, USACGSC RB100-30, vol. 1, US Army Command, (6 August 1976) 43.
57. Ibid., 26.
58. Desmond Ball, 'Nuclear war at Sea', *International Security*, vol. 10, 3 (Winter 1985/86) 3–31.
59. Paul Bracken, *The Command and Control of Nuclear Forces* (New Haven, Conn.: Yale University Press, 1983) 5.
60. Ibid.
61. The standard works on the US intelligence apparatus include: James Bamford, *The Puzzle Palace: Inside the NSA* (Harmondsworth: Penguin Books, 1983); Victor Marchetti and John D. Marks, *The CIA and the Cult of Intelligence* (New York: Dell Books, 1974); and Jeffrey Richelson, *The US Intelligence Community* (Cambridge, Mass.: Ballinger Press, 1985).
62. Jeffrey Richelson, ibid., 80.
63. William Arkin and Richard Fieldhouse, 'Nuclear Weapons Command, Control and Communications', in *SIPRI Yearbook 1984* (London and Philadelphia: Taylor and Francis, 1984) 470.
64. Bruce Blair, op. cit., 1985, 146–7.
65. Daniel Ford, *The Button* (New York: Simon & Schuster, 1987) 61.
66. Bruce Blair, op. cit., 223.
67. Craig Covault, 'New Missile Warning Satellite to be Launched on First Titan 4', *Aviation Week and Space Technology* (20 February 1989) 34–40.
68. Daniel Ford, op. cit., 58–94.
69. 'North Warning Radar', in *C^3I Handbook Vol 1* (Palo Alto, California: EW Publications, 1986) 107–8.
70. Bruce Blair, op. cit., 223.
71. 'Raytheon to Install Phased-Array radar at BMEWS Site in England', *Aviation Week and Space Technology* (12 September 1988) 131.
72. Bruce Blair, op. cit., 261–4.
73. Author's Interview
74. Jakob Hartog, 'The Development of an Air Command and Control System for NATO's Air Forces in Europe', *NATO's Sixteen Nations* (October 1985) 76–8.
75. See: 'NATO Air Defence Ground Environment', in *C^3I Handbook Vol 1*, op. cit., 124–7; Ted Hooton, 'NADGE Enhancement Lay Ground Work for ACCS', *International Defence Review*, 7 (1988) 781–3; and, 'The New NADGE', *International Defence Review*, 5 (1985) 755–8.

76. C.J. Rauch and W.E. Edgington, 'NATO Develops a Long-Term Air Command and Control Plan', *Defence Systems Review* (January 1984) 12–17.
77. Geoffrey Manners, 'NATO's Aerial Sentries on Skywatch Duty', *Jane's Defence Weekly* (24 March 1984) 452–3.
78. Gary Yerkey, 'Greek Participation Will Aid NATO Surveillance Program', *Christian Science Monitor* (17 September 1985) 13.
79. *NATO Facts and Figures*, 1989, op. cit., 345.
80. 'NATO Air Defence Ground Environment', op. cit., 124–7.
81. Jakob Hartog, 'ACCS: A Challenge and an Opportunity', *Signal*, December 1983, 72–4.
82. Jakob Hartog, 'The Development of a NATO Air Command and Control System', *Signal*, October 1984, 51–5.
83. Douglas Barrie, 'NATO Disbands ACCS Team', *Jane's Defence Weekly* (19 August 1989) 287.
84. Douglas Barrie, 'NATO Rethinks Future of ACCS', *Jane's Defence Weekly* (16 June 1990) 167.
85. Daniel Charles, *Nuclear Planning in NATO* (Cambridge, Mass.: Ballinger Press, 1987) 44.
86. Author's Interview.
87. Bruce Blair, 'Alerting in Crisis and Conventional War', in Ashton Carter, John Steinbruner and Charles Zracket (eds), op. cit., 77n.
88. Daniel Charles, op. cit., 40.
89. Kurt Gotfried and Bruce Blair (eds), *Crisis Stability and Nuclear War* (New York: Oxford University Press, 1988) 235.
90. Ibid., 243.
91. The alert is well documented in: Richard M. Nixon, *The Memoirs of Richard Nixon, Vol II* (Warner Books, USA, 1979), 498–500; and, Henry Kissinger, *Years of Upheaval* (London: Weidenfeld & Nicolson, 1982) 587–9.
92. See Scott D. Sagan, 'Nuclear Alerts and Crisis Management', *International Security*, vol. 9, (Spring 1985) 99–139; and Duncan Campbell, *Unsinkable Aircraft Carrier* (London: Michael Joseph, 1985) 316–18.
93. Kissinger himself refers to 'the indisputable fact that there had been no prior consultation over [an] alert that involved US troops stationed in Europe'. See Kissinger, op. cit., 712–14.
94. Kurt Gotfried and Bruce Blair, op. cit., 235–44.

3 US Nuclear Command and Control in NATO

1. Congressional Research Service, *Authority to Order the Use of Nuclear Weapons*, Paper Prepared for the Committee on International Relations, Subcommittee on International Security and Scientific Affairs (Washington DC: US Congress, 1 December 1975) 1.
2. This takes us into areas outside the purview of this thesis. However, the questions can be followed up in: J.M. Smith and C.P. Cotter, *The Powers of the President During Crises* (Washington DC: Public Affairs

Press, 1960); Dorothy James, *The Contemporary Presidency* (Indianapolis: Bobbs-Merrill Co, 1974); and Louis W. Koenig, *The Chief Executive* (New York: HBJ Press, 1975).

3. For an outstanding treatment of this aspect see: Frank G. Klotz, *The US President and the Control of Strategic Nuclear Weapons*, Unpublished PhD Thesis, Oxford University, 1980.
4. Center for Defense Information, *Command, Control and Communications Part I: The Command Structure and Early Warning System*, CDI Briefing Papers, 26 September 1986, 1.
5. Paul Bracken, *The Command and Control of Nuclear Forces* (New Haven, Conn.: Yale University Press, 1983) 123.
6. Peter Stein and Peter Feaver, *Assuring Control of Nuclear Weapons*, CSIA Occasional Paper No 2, Harvard University, 1987, 33.
7. John Guthrie, 'The Doomsday Trigger', *International Combat Arms* (July 1986) 92–3.
8. The Emergency Broadcast System was the means to allow the President to address the nation through television and radio in the event of a serious crisis.
9. Duncan Campbell, op. cit., 1985.
10. Daniel Ford, op. cit., 89–90.
11. Probably the best analysis of the difficulties of using the 'football' can be found in: Bill Gulley and Mary-Ellen Reese, *Breaking Cover* (New York: Simon & Schuster, 1980) 187–90.
12. Center for Defense Information, op. cit., 2.
13. When President Reagan was shot in March 1982, Secretary of State Alexander Haig (a former SACEUR), assumed authority in the White House telling the assembled press: 'as of now, I'm in charge'. In fact, with George Bush, the then Vice President, out of Washington the presidential succession should have passed to the Speaker of the House of Representatives and then to the President *pro tempore* of the Senate before Haig. Moreover Haig, as Secretary of State, was not in the NCA succession chain which would have passed to Secretary of Defense Weinberger (see Daniel Ford, op. cit., 145–6). Alexander Haig is himself commendably candid in his autobiography which relates the incident. He writes that he said:

    > constitutionally . . . you have the President, the V.P., and the Secretary of State in that order . . . As of now, I am in control here, in the White House.

    He goes on, however, to claim that in the melée following the Reagan assassination attempt he made this statement to calm the situation. What is obvious from the Haig account is that no-one of any authority was clear about the delegated pathways of succession beyond the Vice President. See Alexander Haig, *Caveat* (London: Weidenfeld & Nicolson, 1984). The quote is from page 160, but the chapter repays reading, 141–66.
14. Barry Schneider, *Who Leads America if They Destroy Washington?* (Washington DC: National Institute for Public Policy, Information Series No 143, May 1983).

15. *First Use of Nuclear Weapons: Preserving Responsible Control*, Hearing before the House Committee on International Relations, Subcommittee of International Security and Scientific Affairs, 94th Congress, GPO, 1976, 55.
16. Daniel Ellsberg, 'Ellsberg: Who has Finger on US Nuclear Trigger', *Omaha World Herald* (30 August 1981) 1.
17. Bruce Blair, op. cit., 1985, 113.
18. *Authority to Order the Use of Nuclear Weapons*, op. cit., 5–6.
19. Quoted in Simon Duke, *US Defence Bases in the UK* (London: Macmillan, 1987, 39.
20. David Gates, 'American Strategic Bases in Britain: The Agreements Governing Their Use', *Comparative Strategy*, vol. 8 (1989) 103–4.
21. Quoted in Simon Duke, op. cit., 80.
22. John Baylis, 'American Bases in Britain The Truman-Attlee Understandings', *The World Today*, August/September 1985, 155–9.
23. Peter Osnos, 'British say they can veto cruise missile firing', *The Washington Post* (1 November 1983) 1.
24. Daniel Ford, op. cit., 125.
25. Center for Defense Information, op. cit., 3.
26. Ibid.
27. Walter Pincus, 'President's Command Jet Shifted Inland', *The Washington Post* (23 September 1983) 3.
28. 'E-4B Airborne Command Post', in *Jane's Military Communications 1988* (London: Jane's Publishers, 1988) 745–6.
29. Desmond Ball, *Can Nuclear War be Controlled?*, Adelphi Paper 169, IISS, (Autumn 1981) 16–17.
30. Katie Cutler, 'Inside the Looking Glass', *Airman* (September 1982) 7–12.
31. 'EC-135 Post Attack Command and Control Aircraft/Worldwide Airborne Command Post', in *C³I Handbook* (Palo Alto, California: EW Publications, 1986) 117–18. PACCS continuous airborne alert ended in 199O. See James J. Wirtz, 'Ground Alert for Looking Glass: SAC's New Emphasis on Strategic Warning', *Defense Analysis*, vol. 7, 1 (1991) 104–7.
32. Kent Black and Andrew Lindstrom, 'TACAMO', *Signal* (September 1978) 6–11.
33. Michael Donnheim, 'E-6A TACAMO Tail Damage Prompts Delivery Delay', *Aviation Week and Space Technology* (17 April 1989) 28–9. TACAMO aircraft moved from airborne to ground alert in March 1991. See Barbara Starr, 'TACAMO Airborne Alert Abandoned', *Jane's Defence Weekly* (30 March 1991) 461.
34. Nancy Foster, 'Citizens Jam Nuclear Radio Network', *Bulletin of the Atomic Scientists* (November 1988) 21–6.
35. Center for Defense information, op. cit., 5.
36. William Arkin and Richard Fieldhouse, op. cit., 455–516.
37. Desmond Ball (1981), op. cit., 3.
38. The MITRE Corporation, *MITRE: The First Twenty Years* (Bedford, Mass.: MITRE Corporation, 1979) 80.
39. Joseph Rose, 'AUTOVON and AUTODIN: Past Present and Future', *Signal* (April 1969) 12–5.

40. Richard Reaser and Robert Wood, 'AUTODIN is On', *Signal* (April 1970) 6–9.
41. Clarence E. McKnight, 'The Army's National Role in NATO Communications', *Signal* (December 1980) 48.
42. Joe Jones, 'Defense Communications System AUTODIN', *Signal* (September 1969) 74–5.
43. 'Automatic Secure Voice Communications (AUTSECVON)', in *Jane's Military Communications 1988*, op. cit., 744.
44. 'Defense Switched Network', in ibid., 745.
45. The ARPANET is explained fully in Defense Advanced Research Projects Agency (DARPA), *A History of the ARPANET* (Washington DC: DARPA Publications, 1 April 1981).
46. 'Defense Data Network', in *Jane's Military Communications 1988*, op. cit., 744.
47. R.T. Hanson and K. Straussman, 'The US Armed Forces European Telephone System', *Signal* (October 1986) 61–3.
48. Clarence E. McKnight, op. cit., 45.
49. USAF, *Digital European Backbone Facility*, FY1985, Military Construction Project Data, Project No 11470, 18 September 1984.
50. Duncan Campbell (1985), op. cit., 187–8.
51. Ashton Carter, 'Communications Technologies and Vulnerabilities', in Ashton Carter, John Steinbruner and Charles Zracket (eds) *Managing Nuclear Operations* (Washington DC: Brookings Institution, 1987) 247–9.
52. William Arkin and Richard Fieldhouse, *Nuclear Battlefields* (Cambridge, Mass.: Ballinger Press, 1985) Appendix A. This Appendix is a list of overseas US nuclear infrastructure facilities and probably the best single source of information on US facilities in Europe.
53. S. Shiveley, 'A Local Area Network for NATO Headquarters', *Signal* (May 1984) 137–52.
54. Duncan Campbell (1985), op. cit., 49. Arkin and Fieldhouse give 1966 ('Nuclear Weapons Command, Control and Communications', op. cit., 465.
55. Center for Defense Information, op. cit., 5.
56. *MITRE: The First Twenty Years*, op. cit., 177.
57. R.B. Harrison, 'WWMCCS ADP: The Performance Story', *Signal* (January 1985) 47.
58. 'World Wide Military Command and Control System' in *C^3I Handbook*, op. cit., 122–3.
59. R. B. Harrison, op. cit., 47–53.
60. James B. Schultz, 'Communications Connectivity Through Space' in *C^3I Handbook*, op. cit., 169–72.
61. Deputy Under Secretary of Defense for Command Control and Communications, *Modernisation of the WWMCCS Information System (WIS)*, Prepared for the Committee on Armed services, House of Representatives, No 96-916 and 97-333, 31 July 1982. See also David Boutacoff, 'WWMCCS Evolves to Meet Expanding Requirements', *Defense Electronics* (4 November 1984) 79–85.
62. William Arkin and Richard Fieldhouse (1985), op. cit., Appendix A.

63. Duncan Campbell (1985), op. cit., 59.
64. There are differences between the numbers of GIANT TALK/Scope Signal III sites listed by the sources of notes 62 and 63. Arkin and Fieldhouse (1984, 481), state that there were 12 Scope Signal III sites in the northern hemisphere. However, Campbell (1985, 68n) lists 16 GIANT TALK sites which include only two of those mentioned [in my text] in Europe and only one of which (Ascension Island) is in the southern hemisphere.
65. William Arkin and Richard Fieldhouse (1985), op. cit., Appendix A.
66. Duncan Campbell (1985), op. cit., 224.
67. William Arkin and Richard Fieldhouse (1985), op. cit., Appendix A.
68. High Lucas, 'US Radio Link to Submarines at 400ft Depth in Mediterranean', *Jane's Defence Weekly* (21 July 1984) 52.
69. Peter Stein and Peter Feaver, op. cit., 56.
70. Dan Caldwell, *Permissive Action Links: A Description and a Proposal*, Conference on Rethinking the Nuclear Weapons Dilemma in Europe (Minneapolis: University of Minnesota, May 1985) 3.
71. T. Cochran, W. Arkin and M. Hoenig, *Nuclear Weapons Databook Vol. I: US Nuclear Forces and Capabilities* (Cambridge, Mass.: Ballinger Press, 1984) 30.
72. Peter Stein and Peter Feaver, op. cit., 57.
73. Catherine Kelleher, 'NATO Nuclear Operations', in Ashton Carter, John Steinbruner and Charles Zracket (eds), op. cit., 457.
74. William Arkin and Richard Fieldhouse (1985), op. cit., Appendix A.
75. JAYCOR, *Theatre Nuclear Forces Survivability and Security: An Issue Evaluation Plan for Fixed Storage Sites*, DNA 5082F, Prepared for the Defense Nuclear Agency, October 1979, 20. See also Strategic Studies Center, *Increasing the Survivability of TNF Weapons in NATO*, SSC-TN-6070-2, April 1978. I am indebted to Daniel Charles for providing me with these documents.
76. JAYCOR, op. cit., 21.
77. Catherine Kelleher, op. cit., 453.
78. Daniel Charles, *Nuclear Planning in NATO* (Cambridge, Mass.: Ballinger Press, 1985) 57.
79. Catherine Kelleher, op. cit., 455.
80. House of Representatives, Committee on Armed services, Hearing on Military Posture, FY 1981, vol. 2, US Congress 1980, 2329.
81. William Arkin and Richard Fieldhouse (1985), op. cit., Appendix A. The exact location of ECCCS switches are not public knowledge.
82. Kurt Gottfried and Bruce Blair, *Crisis Stability and Nuclear War* (New York: Oxford University Press, 1987) 238.
83. 'Cemetery Net Upgrade (EUCOM C3)', House of Representatives, FY1981, Appropriations Committee, DOD Authorisations, 12 September 1979, 748.
84. Quoted in Duncan Campbell (1985), op. cit., 184.
85. Regency Net seems to have assumed some of the peacetime role of the ECCCS.
86. House of Representatives, 'Regency Net', Subcommittee on Appropriations, DOD Authorisation, FY 1984, 347.

87. 'Regency Net Contract', *Signal* (April 1984) 72.
88. US Army, *Operations*, FM100-5, July 1976.
89. Catherine Kelleher, op. cit., 457.
90. See, for example Daniel Charles, op. cit., 134–5.
91. The 'top-down' and 'bottom-up' release chains are discussed in Catherine Kelleher, op. cit., 457–62.
92. US Senate Committee on Armed Services, DOD Appropriations, FY1983, 97th Congress, 4334.
93. D. Hingorani and R. Brand, 'Architectural Framework for the Evolution of NATO Integrated Communications System', *Signal* (October 1985) 56.
94. Larry K. Wenz, 'NATO/National Strategic Defense Communications Systems: Interoperability and Rationalisation – Rhetoric or Reality', *Signal* (October 1985) 86.
95. Ibid.

4 Command and Control of British Nuclear Weapons

1. The text of the Nassau Agreement is reproduced in Andrew J. Pierre, *Nuclear Politics: The British Experience With An Independent Strategic Force, 1939–70* (Oxford: Oxford University Press, 1972) 346–7.
2. Congressional Research Service, *Authority to Order the Use of Nuclear Weapons* (Washington DC: Library of Congress, 1 December 1975) 11.
3. Max Hastings and Simon Jenkins, *The Battle for the Falklands* (London: Michael Joseph Ltd, 1983) 81.
4. Duncan Campbell and Patrick Forbes, 'New Whitehall Bunker', *New Statesman*, 26 July 1985, 6.
5. Paul Brown, 'MoD Digs Deep for Bunker', The *Guardian*, 25 July 1985, 4.
6. Duncan Campbell and Patrick Forbes, op. cit., 6.
7. Duncan Campbell, 'Maggie's Bunker', *New Statesman*, 24 September 1982, 6–8.
8. Ibid., 6.
9. Ibid., 7.
10. Duncan Campbell, *War Plan UK* (London: Paladin Books, 1982) 217.
11. Author's Interview.
12. Author's Interview.
13. Lawrence Freedman, Britain and Nuclear Weapons (London: Macmillan Press, 1980) xiii.
14. Benjamin Schemmer, 'Who Would Start a War?', *Armed Forces Journal International* (December 1981) 36–46.
15. Simon Freeman and Barrie Penrose, 'How Britain Would Have Coped if the Cabinet Had Been Killed', *The Times*, 21 October 1984, 10.
16. Dan Caldwell, *PALs for Allies and Adversaries*, paper prepared for the Conference on Rethinking the Nuclear Weapons Dilemma in Europe, University of Minnesota, Winsconsin, May 2–5 1985, 5.
17. Author's Interview.
18. Lawrence Freedman (1980), op. cit., xiii.
19. Ibid.

20. A. J. R. Groom, *British Thinking About Nuclear Weapons* (London: Frances Pinter, 1974) 513.
21. W. T. T. Pakenham, *Naval Command and Control* (London: Brassey's Defence Publishers, 1989) 107–8.
22. Duncan Campbell (1982), op. cit., 232.
23. P. M. Ford, 'The Defence Communications Network', *Journal of the Royal Signals Institute* (Summer 1989) 68–74.
24. The BACKBONE network is described in Campbell (1982), op. cit., 240–8.
25. Mark Urban, 'Fibre Optics are Arteries of Information Network', The *Independent*, 12 January 1989, 5.
26. Mark Urban, 'P.M. to Get High-Tech Command Centre', The *Independent*, 12 January 1989, 1.
27. Duncan Campbell, 'The Parliamentary Bypass Operation', *New Statesman*, 23 January 1987, 8–10.
28. J. T. Richelson and Desmond Ball, *The Ties that Bind* (London: Allen and Unwin, 1985) 13–28.
29. Duncan Campbell (1982), op. cit., 227, and 'The Men who Wait for Nuclear Attack', The *Guardian*, 27 October 1980, 4. These two sources refer to High Wycombe. The evidence for Northwood relates to an interview in which I was assured that the Royal Navy was tied in to all early warning systems indepedently of the RAF.
30. Daniel Ford, *The Button* (London: Allen and Unwin, 1985) 61.
31. 'UK to update Missile Detection System', *Jane's Defence Weekly*, 7 June 1986, 1028. See also William Arkin and Richard Fieldhouse, 'Nuclear Weapons Command and Control', in *SIPRI Yearbook 1984* (London & Philadelphia: Taylor and Francis, 1984) 470.
32. 'The Men who Wait for Nuclear Attack', op. cit.
33. Ibid.
34. Duncan Campbell (1982), op. cit., 213 and Author's Interview.
35. Missile flight times vary enormously depending upon the location from which the missiles are fired and their trajectory. The range is typically between 5 and 15 minutes within the European theatre. Ultimately warning depends not simply on flight times but also on the speed with which radar and other early warning assets can detect, confirm and transmit the missile launch information to the relevant command centres.
36. Bruce Blair, *Strategic Nuclear Command and Control* (Washington DC: Brookings Institution, 1985) 223.
37. Duncan Campbell (1982), op. cit., 228.
38. 'North Warning Radar', in *Command, Control and Communications Handbook* (Palo Alto, California: EW Publications, 1986) 107.
39. 'RAF Sets BMEWS Test Timetable', *Jane's Defence Weekly*, 12 May 1990, 883.
40. 'Britain's Fence Takes Shape', *Jane's Defence Weekly*, 15 December 1984, 1078–80; 'Martello – Modern 3-D Surveilllance Radar', *GEC Journal of Research*, vol. 3, 2 (1985) 104–13; and 'Closing NATO's Backdoor', *Interavia*, vol. 2 (1983) 13.
41. Mark Hewish, 'Britain's Air Defence Grows New Teeth', *New Scientist*, 12 March 1981, 684.

42. 'Shackleton is Better than Nothing says C-in-C', *Jane's Defence Weekly*, 30 November 1985, 1182–3.
43. 'Britain's Fence Takes Shape', op. cit., 1079.
44. 'New AD Sensors for the UK', *Defence Attaché*, 2, (1985) 6.
45. Duncan Campbell, 'New Spy Station to Look Inside USSR', *New Statesman*, 21–28 December 1984, 6.
46. 'New AD Sensors for the UK', op. cit., 6.
47. Paul Brown and David Fairhall, 'Anger over Plan for radar Station at Beauty Spot', The *Guardian*, 5 April 1990, 24.
48. Malcolm Spaven, 'ELF – Surviving the Traumas Part II', *Jane's Defence Weekly*, 30 November 1985, 1196.
49. 'Two Important Naval Communications Contracts', *Defence Materiel*, January–February 1985, 19.
50. Ibid.
51. Malcolm Spaven, 'Communicating with Submarines', *Jane's Defence Weekly*, 23 November 1985, 1154.
52. Ibid.
53. Duncan Campbell, 'The Deterrent Goes to War', *New Statesman*, 1 May 1981, 9.
54. 'Revealed, US Role in UK SSBN Communications Link', *Defence*, 1 (January 1987) 6.
55. Bruce Blair, op. cit., 271.
56. Ibid., 270.
57. Written Answers, *Hansard*, 14 February 1991, column 592. A later Written Answer confirmed that the ELF project would not be deployed elsewhere, see Hansard, 8 March 1991, column 331.
58. *Strategic Nuclear Weapons Policy*, House of Commons, Defence Committee, Session 1980–1, 87 (para 331).
59. Malcolm Spaven, 'ELF – Surviving the Traumas Part II', op. cit., 1196.
60. Bruce Blair, op. cit., 119.
61. T. W. Pakenham, op. cit., 86–90.
62. Bruce Blair, op. cit., 271.
63. Desmond Ball, *Targeting for Strategic Deterrence*, Adelphi Paper 1985, IISS, 1983, 9 *et passim*.
64. 'The Nerve Centre of US Nuclear Strategy', *NATO Letter*, May 1966, 19–21.
65. Lawrence Freedman, 'British Nuclear Targeting', *Defence Analysis*, vol. 1, 2 (1985) 90.
66. Jonathan Crane, *Submarine* (London: BBC Books, 1984) 198–200.
67. Author's Interview.
68. Author's Interview.
69. Lawrence Freedman (1980), op. cit., xiii.
70. Peter Jackson, *Strike Command* (London: Ian Allen, 1984) 12.
71. Terry Gander, 'RAF Germany', in *The Modern Royal Air Force* (Cambridge, UK: Patrick Stephens Limited Publications, 1984) 29.
72. Author's Interview.
73. Author's Interview.
74. 'Jaguar Mission', *RAF News*, 9–23 September 1983, 11.
75. Ibid.

76. 'RAF Honington', *RAF News*, 16 December 1982, 12–13.
77. Peter Jackson, op. cit., 8.
78. 'RAF Developments', *Defence Attaché*, 3 (1981) 30.
79. 'British Developments in Satellite Communications', *International Defence Review*, 5 (1985) 722.
80. Paul Beaver, *Modern Royal Navy* (Cambridge, UK: Patrick Stevens Publishers, 1982) 19.
81. R. Pengelley, 'OPCON 2 – a Strategic maritime C2 System', *Defence Attaché*, 6 (1981) 9.
82. Malcolm Spaven, *Royal Navy Nuclear Capable Ships*, Unpublished Briefing, ADIU, University of Sussex, 1985.
83. Duncan Campbell, 'Too Few Bombs To Go Round', *New Statesman*, 29 November 1985, 12.
84. Author's Interview.
85. William Arkin and Andrew Burrows, *British Nuclear Weapons In The Falklands*, Unpublished Paper, (Washington DC: IPS, 1984).
86. R. Pengellet, op. cit., 9–15; W. T. T. Pakenham, op. cit., 108; and R. J. Raggett, 'Naval Communications Experience – the Good and Bad Aspects of Technology', *Maritime Defence*, August 1979, 312–3.
87. F. J. Schultz, 'C3 in NATO', in Kaltefleiter and Schumacher (eds) *Conflicts, Options, Strategies in a Threatening World* (Kiel Christian-Albrechts University, Institute of Political Sciences, 1982) 208–18.
88. *Jane's Military Communications 1988* (London: Jane's Defence Publishers, 1988) 741.
89. This description of the Lance Missile system and firing operations has been put together from the following sources: R. A. Spackman, 'Lance: Field Artillery Missile System', *Journal of the Royal Artillery*, September 1972, 127–33 and D. Richardson, 'Lance into Battle', *Flight International*, 30 April 1977, 1192–5.
90. A. J. Pierre, op. cit., 167.
91. Peter Malone, *The British Nuclear Deterrent* (London: Croom Helm, 1984) 95.
92. A. J. R. Groom, *British Thinking About Nuclear Weapons* (London: Frances Pinter, 1974) 513.
93. Lawrence Freedman, 'The Role of Third-Country Nuclear Forces', in J. D. Boutwell, P. Doty, and G. Treverton (eds) *The Nuclear Confrontation in Europe* (London: Croom Helm, 1985) 121.
94. Peter Malone, op. cit., 94.
95. Lawrence Freedman, op. cit., 121.

5 French Nuclear Command and Control

1. 'L'engagement nucléaire ne peut reposer que sur la décision d'un seul', *Le Monde*, 19 November 1980, 12.
2. M. Schneider, 'Une monarchie nucléaire', *Le Monde*, 8 December 1983, 13.
3. Discussion of this decree can be found in R. Girardet, *Problèmes Contemporains de Défense Nationale* (Paris: Dalloz, 1974) 160–1.

4. 'M. Mauroy: un ordre d'engagement de l'arme nucléaire est peu compatible avec une concertation préalable', *Le Monde*, 11 March 1984, 11.
5. 'L'engagement nucléaire ne peut reposer que sur la décision d'un seul', op. cit., 12.
6. J. Isnard, 'Le code d'engagement de la force nucléaire', *Le Monde*, 20 May 1981, 12.
7. Ibid.
8. An enabling code means a code that must be inserted into the nuclear weapon to 'unlock' it, thereby enabling the launch, firing or detonation of the weapon to occur. Without the code the weapon is inert and militarily useless. An authenticating code is one which identifies and validates the source of an order to use nuclear weapons (or carry out some other operation) but does not physically affect the readiness of the weapon for use.
9. J. Isnard, op. cit.
10. W.L. Kohl, *French Nuclear Diplomacy* (Princeton, N.J.: Princeton University Press, 1971) 182.
11. C. Kelleher, 'NATO nuclear operations' in A. Carter, J. D. Steinbruner and C. Zracket (eds) *Managing Nuclear Operations* (Washington DC: Brookings Institution, 1987) 468.
12. In the mid-1970s strong speculation arose that President Giscard d'Estaing had predelegated release of tactical nuclear weapons to the service chiefs of staff. Under President Mitterrand the 1983 'loi de programmation' ended this speculation by pointedly taking nuclear command authority from the service chiefs and placing it in the hands of the Chief of the General Staff under the direct authority of the President (author's interview).
13. See D. Yost, 'French nuclear targeting' in D. Ball and J. Richelson (eds) *Strategic Nuclear Targeting* (Ithaca, N.Y.: Cornell University Press, 1986) 127–56.
14. J. Isnard, op. cit.
15. R. Smith, 'Fingers on the French button', The *Observer*, 3 July 1983, 6.
16. 'Inside the Elysée', *The Sunday Times* (Magazine), 30 July 1985, 29.
17. J. Isnard, op. cit.
18. Ibid.
19. D. Yost, *France's Deterrent Posture and Security In Europe Part I: Capabilities and Doctrine*, Adelphi Paper 194 (London: IISS, Winter 1984/5) 26.
20. A. Schwartzbrod and M.J. Gething, 'French air power – the turning point', *Defence*, September 1988, 645.
21. W. Arkin and R. Fieldhouse, *Nuclear Battlefields* Appendix D (Cambridge, Mass.: Ballinger Press, 1985) 285.
22. 'The French Armaments Industry in 1989', *Defence*, March 1989, 194.
23. 'The future of French aerospace', *Defense & Armament Héraclès International*, 77 (October 1988) 51–2.
24. P. Denarie, 'Helios and Syracuse: military space at technospace exhibition', *Defense & Armament Héraclès International*, 81 (February 1989) 57.

25. M. Faivre, 'Le budget de la défense pour 1989', *Défense Nationale* (January 1989) 173.
26. 'The French armament industry in 1989', op. cit.
27. M. Faivre, 'Le renseignement du champ de bataille', *Défense Nationale* (August–September 1988) 176.
28. *Jane's Weapons Systems 1987–88* (London: Jane's Defence Publishers, 1988) 222–3.
29. 'France buys fourth E-3', *Defence* (December 1987) 606.
30. R. Tourrain, *Rapport d'Information par la Commission de la Défense Nationale at des Forces Armées sur l'état et la modernisation des forces nucléaires francaises*, 1730, (Paris: Assemblée Nationale, October, 1980) 268.
31. P. Henin, 'Les télécommunications militaires', *Défense Nationale* (May 1987) 144–5.
32. G.D. Hingorani and R. Brand, 'Architectural framework for the evolution of NATO integrated communications system', *Signal* (October 1985) 58.
33. 'RAMSES nuclear order network delivered', *Jane's Defence Weekly*, 30 April 1988, 849.
34. 'RAMSES communications network completed', *Maritime Defence* (May 1988) 196.
35. M. Faivre (1989), op. cit., 173.
36. These include Cazaux and Mont-de-Marsan which are permanent bases, and Cambrai, Creil, Istres, Luxeuil and Orange which are dispersal bases. See Arkin and Fieldhouse, op. cit., 282–7.
37. Ibid.
38. David Yost (1984–5), op. cit., 26.
39. 'Les Etats-Unis acceptant de vendre à la France des émetteurs pour sa force dissuasion', *Le Monde*, 16 September 1982, 14.
40. 'French nuclear command post enters operational service', *Aviation Week and Space Technology*, 8 February 1988, 253.
41. 'Aerial command post', *Jane's Defence Weekly*, 13 February 1988, 253.
42. 'L'espace pour voir, écouter, communiquer', *Défense Nationale*, July 1986, 146.
43. *Jane's Military Communications 1988* (New York: Jane's Defence Publishers, 1988) 322.
44. P. Denarie, op. cit., 57.
45. A. Schwartzbrod, 'French Navy in the Gulf', *Defence* (September 1987) 505.
46. J.M. Leonorvitz, 'Attitude lost on France's Telecom 1B', *Aviation Week and Space Technology*, 25 January 1988, 30.
47. 'The Syracuse II programme', *Defense & Armament Héraclès International* (December 1988) 72.
48. Plus one in reserve.
49. P. Denarie, op. cit., 57.
50. *Jane's Weapons Systems 1988-89, (London & New York: Jane's Defence Publishers, 1989) 119–20.*
51. P. Chilton, 'French Nuclear Weapons', in J. Howorth and P. Chilton (eds) *Defence and Dissent in Contemporary France* (London: Croom Helm, 1984) 147.

52. *Jane's Weapons Systems 1988-89*, op. cit., 119.
53. P. Chilton, op. cit., 147.
54. J. Deygout, 'RITA: a modern response to battlefield communications needs', *Signal* (March 1984) 26–33.
55. 'Advanced RITA for Hades missiles launch system', *Jane's Defence Weekly*, 10 October 1987, 801.
56. 'The Syracuse II programme', op. cit., 72.
57. P. Chilton, 148.
58. J. Isnard, 'France to update nuclear arsenal with ASMP', *Jane's Defence Weekly*, 30 July 1988, 6.
59. 'New HF radio links for French navy', *Maritime Defence*, September 1986, 11.
60. David Yost (1986), op. cit., 127–56.
61. M. Ducal, 'L'Arme Nucléaire: pour quoi faire?', *Défense Nationale*, February 1989, 17–19.
62. F. Maurin, 'Entretien avec le Général Maurin', *Défense Nationale* (July 1974) 17.
63. J. Lacaze, 'La politique militaire', *Défense Nationale* (November 1981) 15.
64. I. Margine, 'L'Avenir de la dissuasion', *Defénse Nationale* (April 1978) 29.
65. C. Hernu, 'Répondre aux défis d'un monde dangereux', *Défense Nationale* (December 1981) 16.
66. See J.B. Margeride, 'Réflexions sur les systèmes à l'ère électro-informatique', *Stratégique* (Winter 1984) 115–17; and M. Faivre, 'Menaces et risques majeurs pour la sécurité de la France', *Défense Nationale* (December 1988) 22.
67. *Jane's Weapons Systems 1988–89*, op. cit., 4.
68. W.L. Kohl, op. cit., 182.
69. David Yost (1986), op. cit., 139.
70. W. Arkin *et al.*, 'Nuclear weapons', in *SIPRI Yearbook 1988* (New York: Oxford University Press) 51–2.
71. *Jane's Weapons Systems 1988–89*, op. cit., 4.
72. The lower figure derives from 3x16 single warhead M-20 missiles and the upper from 2x16 six-warhead M-4 missiles and 3x16 single warhead M-20 missiles.
73. David Yost (1984–5), op. cit., 26.
74. J. Isnard, 'Pure deterrence – or tainted strategy', *Defense & Armament Héraclès International* (June 1988) 31.
75. R.W. Stafford, 'Defense planning in NATO: a consensual decision-making process', in R. L. Pfaltzgraff and U. Ra'anan (eds) *National Security Policy: the Decision-making Process* (Hamden, Conn.: Archon Books, 1984) 153–4.
76. *NATO Facts and Figures* (Brussels: NATO Information Service, 1986) 93.
77. The Defence Planning Committee which comprises the NAC minus the French delegation and has special responsibility for defence matters could also have become the principal forum for crisis management, but it is difficult to imagine that NATO would have excluded French

representatives or that the French would have allowed themselves to be excluded.

78. J. Anderson, 'NATO Communications requirements and the role of NICSMA', *Journal of the Royal Signals Institute*, vol. 15, 4 (Spring 1982) 160.
79. 'France and Germany agree to set up "hot-line"', *Christian Science Monitor*, 26 August 1985, 2.
80. Franco-German summit statement, 27-8 February 1986 (excerpts), translated and reproduced in *Survival*, vol. 28, 4 (July–August 1986) 366.
81. R. Ullman, 'The covert French connection', *Foreign Policy* (Summer 1989) 21.
82. Ibid.
83. Interview with General Bernard Rogers, *US News and World Report* (15 June 1981) 25.
84. Conference de presse de Monsieur François Mitterrand à l'issue des manoeuvres Kecker Spatz, Présidence de la République, 24 September 1987.
85. 'The nerve centre of US strategy', *NATO Letter* (May 1966) 19–21.
86. R. Ullman, op. cit., 24.
87. P.J. Sloyan, 'America in secret computer deal with France', *The Guardian*, 5 February 1985, 11.
88. R. Ullman, op. cit., 24–6.
89. N. Ashford and J. Eisenhammer, 'A Cartesian guide to defending France', *The Independent*, 28 January 1988, 5.
90. J. Boyes, 'BICES: linking NATO intelligence', *Signal* (October 1987) 19–20.
91. C. Rauch and W. Edgington, 'NATO develops a long-term air command and control plan', *Defense Systems Review* (January 1984) 12–17.
92. L.K. Wentz, 'Outlook for NATO communications', *Signal* (December 1982) 55.
93. G. Turbe, 'Testing Franco-German military co-operation', *International Defence Review*, 11 (1987) 1458.
94. 'SINTAC – the French answer to JTIDS', *Defence* (December 1987) 630–1.
95. *Jane's Military Communications 1988*, op. cit., 331.
96. 'Avant-garde avionique', *Defence* (January 1988) 38–9.
97. J. Isnard, 'France and Europe's security', *Defense & Armament Héraclès International* (November 1987) 71.
98. R. Ullman, op. cit., 26.
99. J.M. Legge, *Theatre Nuclear Weapons and the NATO Strategy of Flexible Response* (Santa Monica, California: RAND Corporation, April 1983) 26–7.

6 Political Decision-Making in Crisis and War

1. Hans Morgenthau, 'Alliances in Theory and Practice', in Arnold Wolfers, *Alliance Politics in the Cold War* (Baltimore, Md.: Johns Hopkins Press, 1959) 191.

2. George F. Liska, *Nations in Alliance: The Limits of Interdependence* (Baltimore, Md.: Johns Hopkins Press, 1962) 175. See also the seminal work of William H. Riker, *The Theory of Political Coalitions* (New Haven, Conn.: Yale University Press, 1962).
3. E. Vandervanter, *Some Fundamentals of NATO Organisation* (Santa Monica, Calif.: RAND Corporation, April 1963) RM-3559-PR, 12.
4. E. W. Boyd *et al.*, *NATO Management: Peace to Crisis Transition* (Santa Monica, Calif.: RAND Corporation, February 1980) R-2576, 12.
5. Andre De Staercke, *NATO's Anxious Birth* (London: Hurst Press, 1985).
6. The text of the North Atlantic Treaty is reproduced in *NATO Facts and Figures* (Brussels: NATO Information Service, 1989) 377.
7. Ibid., 14.
8. Desmond Ball and Jeffrey Richelson, *The Ties that Bind* (London: Allen & Unwin, 1985).
9. Richard K. Betts, *Surprise Attack* (Washington DC: Brookings Institution, 1982).
10. Robert E. Osgood, *Nuclear Control in NATO* (Baltimore, Md.: Johns Hopkins University, 1962).
11. *Communications and Crisis Management in the Alliance*, Western European Union, Paris, 23rd Session, Part II, Document 757, 4 November 1977, 5.
12. Ibid., 7.
13. Frederic L. Kirgis, 'NATO Consultations as a Component of National Decision-Making', *American Journal of International Law*, vol. 73, 3 (July 1979) 395.
14. Lester H. Brune, *The Missile Crisis of October 1962* (Claremont, California: Regina Books) 55.
15. Elie Abel, *The Missile Crisis* (New York: Praeger Books, 1966) 122.
16. Roger Hilsman, *To Move a Nation* (New York: Delta Books, 1967) 212.
17. Frederic L. Kirgis, op. cit., 400.
18. George Ball, *The Past Has Another Pattern* (London and New York: Norton Books, 1982) 298.
19. *Communications and Crisis Management in NATO*, op. cit., 9.
20. *NATO Final Communiqués 1949–74* (Brussels: NATO Information Service, 1974) 188.
21. *Communications and Crisis Management in NATO*, op. cit., 11.
22. Jon McLin, *NATO and the Czechoslovakian Crisis, Part II: Invasion, Reaction and Stocktaking*, Western Europe Series, vol. 4, 4, American University Field Staff, 1969, 3.
23. R. K. Betts, op. cit., 84.
24. Ibid., 85.
25. Kurt Gottfried and Bruce Blair, *Crisis Stability and Nuclear War* (Oxford: Oxford University Press, 1988) 201.
26. Marvin and Bernard Kalb, *Kissinger* (Boston, Mass.: Little, Brown, 1974) 490.
27. Henry Kissinger, *Years of Upheaval* (London: Michael Joseph, 1982) 590.
28. *Communications and Crisis Management in NATO*, op. cit., 13.

29. *NATO Final Communique 1975–80* (Brussels: NATO Information Service, 1980) 153.
30. *Communications and Crisis Management in NATO*, op. cit., 14.
31. 'Crisis Cabinet for NATO', *Atlantic Community News* (March 1980) 2.
32. Benjamin Schemmer, 'Haig Says NATO Can Now Expect 8–14 Days Warning, Not 48 Hours', *Armed Forces Journal International* (October 1977) 16.
33. Barbara Starr, 'NATO Will Get 1 Month Warning of Attack', *Jane's Defence Weekly*, 23 December 1989, 1366–7.
34. Henry Kissinger, op. cit., 713.
35. Arthur Schlesinger, *A Thousand Days* (Boston, Mass.: Houghton Mifflin, 1965) 803.
36. Harland Cleveland, *NATO: The Transatlantic Bargain* (New York: Harper & Row, 1970) 21.
37. Daniel Charles, *Nuclear Planning in NATO* (Cambridge, Mass.: Ballinger Press, 1985) 44.
38. R. K. Betts, Op. cit., 176.
39. The latter refers to the raising of the DEFCON following the assassination attempt on Ronald Reagan. See Alexander Haig, *Caveat* (Weidenfeld & Nicolson, 1984) 156.
40. Daniel Charles, op. cit., 44.
41. Author's Interview.
42. Author's Interview
43. Author's Interview
44. Fred Ikle, 'NATO's First Nuclear Use: A Deepening Trap', *Strategic Review* (Winter 1980) 20.
45. Richard K. Betts, 'Surprise Attack: NATO's Political Vulnerability, *International Security*, vol. 5, 4 (Spring 1981) 134.
46. See for example: J. Critchley *et al.*, 'Can the Use of Nuclear Weapons be Decided Upon Collectively?', *NATO Letter* (October 1965) 18–22.
47. *US Security Issues in Europe: Burden Sharing and Offset; MBFR and Nuclear Weapons*, Staff Report, Committee on Foreign Relations, US Senate, 93rd Congress, 2 December 1973, 19–20.
48. Paul Bracken, 'The NATO Defense Problem' *Orbis* (Spring 1983) 89. See also Paul Bracken, 'Collateral Damage and Theatre Warfare', *Survival*, vol. 22, 5 (Sept–Oct 1980) 203–7; Paul Bracken, 'Urban Sprawl and NATO Defence', *Survival*, vol. 18, 6 (Nov–Dec 1976) 254–60; and, Gary L. Guertner, 'Nuclear War in Suburbia', *Orbis*, vol. 26, 1 (1982) 49–70.
49. Admiral Sir James Eberle, quoted in *Command and Control*, a Channel Four Equinox documentary made by 20/20 Television and broadcast on Thursday 22 October 1987. Transcript and video are available.
50. Tony Catterall, 'War Game "Idiocy" fires Kohls Resolve', The *Observer*, 30 April 1989, 27.
51. 'Der Iwan Kommt – und Feste Druff', *Der Spiegel*, 18 (1989) 23–7. I am grateful to Dr David Stevenson for the translation of this article.
52. Admiral Sir James Eberle, op. cit.
53. 'Der Iwan Kommt – und Feste Druff', op. cit., 26.
54. Joseph D. Douglass, 'Soviet Nuclear Strategy in Europe: A Selective Targeting Doctrine?', *Strategic Review* (Fall 1977) 12–23.

55. Desmond Ball, 'Soviet Strategic Planning and Control of Nuclear War', in Roman Kolkowicz and Ellen Mickiewicz (eds) *The Soviet Calculus of Nuclear War* (Lexington, Mass.: Lexington Books, 1986) 55–6.
56. Ibid., 53.
57. J. M. Legge, *Theatre Nuclear Forces and the NATO Strategy of Flexible Response* (Santa Monica, Calif.: RAND Corporation, April 1983) R-2964-FF, 26.
58. Ibid., 27.
59. Author's Interview.
60. Phil Williams, 'NATO Crisis Management: Dilemmas and Trade-Offs', *The Washington Quarterly*, vol. 12, 2 (Spring 1989) 35.
61. Desmond Ball *et al.*, *Crisis Stability and Nuclear War* (Ithaca, N.Y.: American Academy of Sciences and the Cornell University Peace Studies Program, 1987) 59.
62. Richard K. Betts, Soldiers, *Statesman and Cold War Crises* (Cambridge, Mass.: Harvard University Press, 1977) 216.
63. Catherine Kelleher, op. cit., 458.
64. Richard Ned Lebow, *Between Peace and War: the Nature of International Crises* (Baltimore, Md.: Johns Hopkins Press, 1981) 46.
65. Kurt Gottfried and Bruce Blair, op. cit., 276–7.
66. Scott Sagan, 'Nuclear Alerts and Crisis Management', *International Security* vol. 9, 4 (Spring 1985) 102.
67. 'Test alert attributed to Gates', *The Washington Post*, May 17, 1960, 1.
68. Alexander Haig, op. cit., 156–7.
69. Phil Williams, op. cit., 35.
70. R. K. Betts (1981), op. cit., 131.
71. Kurt Gottfried and Bruce Blair, op. cit., 277.
72. John Steinbruner, 'An Assessment of Nuclear Crises', in F. Griffiths and J. C. Polanyi, *The Dangers of Nuclear War* (Toronto, Ont.: University of Toronto Press, 1979) 38.
73. Lester Brune, op. cit., 63.
74. David Detzer, *The Brink: The Cuban Missile Crisis of 1962* (London: J. M. Dent, 1980) 246.
75. Roger Hilsman, op. cit., 221.
76. R. K. Betts (1981), op. cit., 132.
77. Scott D. Sagan, op. cit., 133.
78. Bruce Blair, 'Alerting in Crisis and Conventional War', in Ashton Carter, John Steinbruner and Charles Zracket (eds) *Managing Nuclear Operations* (Washington, D.C.: Brookings Institution, 1987) 83.
79. Paul Bracken, *The Command and Control of Nuclear Forces* (New Haven, Conn.: Yale University Press, 1983) 59–65.
80. Catherine Kelleher, 'NATO Nuclear Operations', in Carter, Steinbruner and Zracket, op. cit., 453.
81. Jeffrey Record, 'Theatre Nuclear Weapons: Begging the Soviet Union to Pre-empt', *Survival* (September–October 1977) 208–11.
82. Richard Perle, *US Senate Committee on Armed Services 1985*, US GPO, 1985, Pt VII, 3942.
83. Kurt Gottfried and Bruce Blair, op. cit., 238.
84. This point is explained in Bracken (1983), op. cit., 164–169.

85. Ibid.
86. Evidence that the Soviet Union can monitor these communications was provided in 1976 when, during a NATO exercise (probably WINTEX) to test release systems and procedures, the Soviet Union announced NATO's intention to release nuclear weapons several hours before NATO formally ordered the decision. See Charles, op. cit., 145.
87. NATO Information Service, *NATO Facts and Figures* (Brussels: NATO Information Service, 1989) 285–6.
88. Ibid., 282.
89. Joachim Sochaczewski, 'NATO CCIS', *Military Technology*, 10 (1988) 31–7.
90. Joachim Sochaczewski, op. cit., 31–7.
91. Author's correspondence with SHAPE headquarters, 22 November 1989.
92. 'Conflicting Systems Negate NATO C^3 Lead', *C^3I Report*, 25 January 1988, 7.
93. Franz-Joseph Schulze, 'C^3 in NATO', in Werner Kaltefleiter and Ulricke Schumacher (eds) *Conflicts, Options, Strategies in a Threatening World* (Kiel: Christian Albrecht University, Institute of Political Science, 1982) 223.
94. 'ACE-HIGH Transmission System', *Jane's Military Communications 1988* (London: Jane's Publishing Company, 1988) 717.
95. General Richard C. Ellis, in *Seminar on Command, Control, Communications and Intelligence*, Programme on Information Resource Policy, Harvard University, 1982, 2.
96. John H. Cushman, *Command and Control of Theatre Forces: Adequacy* (Cambridge, Mass.: Harvard University Press, 1983) 3.
97. Bob Raggett, 'NATO C^3 in Disarray', *Jane's Defence Weekly*, 8 September 1984, 418.
98. Catherine Kelleher, 'NATO Nuclear Operations', in Carter, Steinbruner and Zracket (eds) op. cit., 468.
99. These are the means selected by Desmond Ball in his assessment of the vulnerability of the US strategic nuclear command and control network. See Desmond Ball, *Can Nuclear War Be Controlled* (London: Adelphi Paper 169, IISS, 1981) 9–14.
100. Desmond Ball (1987), op. cit., 14.
101. Ibid., 15.
102. Clark C. Abt, *A Strategy for Terminating a Nuclear War* (Boulder, Colo.: Westview Press, 1985); Steven Cimbala, *Strategic War Termination* (New York: Praeger, 1986); and, Herman Kahn *et al.*, *War Termination: Issues and Concepts* (New York: Hudson Institute, 1968).
103. Robert G. Leahy, 'C^3 for Strategic Nuclear War termination', *Signal* (August 1988) 40.
104. Paul Bracken, op. cit., 177–8.

Conclusions

1. See Roderic Braithwaite, 'Russian Realities and Western Policy', *Survival*, vol. 36, 3 (Autumn 1994) 11–27; Thomas Goltz, 'The Hidden

Russian Hand', *Foreign Policy*, 92 (Autumn 1993) 92–116; Jacob W. Kipp, 'The Zhirinovsky Threat', *Foreign Affairs*, vol. 73, 3 (May–June 1994) 72–86; Vladmir Lukin, 'Our Security Predicament', *Foreign Policy*, 88 (Autumn 1992) 57–75; and, Dmitri Simes, 'The Return of Russian History', *Foreign Affairs*, vol. 73, 1 (January–February 1994) 72–86.

2. Samuel Huntington, 'The Clash of Civilisations', *Foreign Affairs*, vol. 72, 3 (Summer 1993) 22–49. This type of analysis is now abundant. Amongst the most interesting similar studies are: Robert Jervis, 'The Future of World Politics: Will It Resemble the Past?', *International Security*, vol. 16, 3 (Winter 1991–92) 39–73; James Kurth, 'Things to Come: The Shape of the New World Order', *The National Interest* (Summer 1991) 3–12; and, James Mayall, 'Nationalism and International Security After the Cold War', *Survival*, vol. 34, 1 (Spring 1992) 19–35.
3. Perhaps the best single study of these issues is Malcolm Dando and Paul Rogers, *A Violent Peace: Global Security After the Cold War* (London: Brassey's, 1992).
4. See John Esposito, *The Islamic Threat: Myth or Reality?* (Oxford: Oxford University Press, 1992); Leon T. Hadar and Judith Miller, 'Is Islam a Threat?', *Foreign Affairs*, vol. 72, 1 (Spring 1993) 27–56; and Ghassan Salame, 'Islam and the West', *Foreign Policy*, 90 (Spring 1993), 22–37.
5. For an excellent discussion of the future of the NATO alliance see Martin Smith, *The Future of NATO: Adjustment or Decline?* (London: Macmillan, 1995), [forthcoming].
6. These issues are set out with great clarity in: Michael Quinlan, 'The Future of Nuclear Weapons: Policy for Western Possessors', *International Affairs*, vol. 69, 3 (1993) 485–96.
7. The experience of the Gulf War of 1990–1 should not be taken as evidence that NATO has solved its interoperability problems. The Gulf War was a distinct and particular event in which joint operations were facilitated by keeping national missions distinct, sharing primarily American technology which was handed round for the purpose, and by sustained efforts to build connectivity where none existed. In the words of one senior military commander speaking at the time: '[the allies] put more communications connectivity [into the Gulf theatre] than they have had in Europe during the past forty years'. These issues are explored more fully in Shaun Gregory, *Command, Control, Communications and Intelligence in the Gulf War* (Australian National University, Strategic and Defence Studies Centre, September 1991), Working Paper No 238.

Bibliography

Books and Monographs

Elie Abel, *The Missile Crisis* (New York: Praeger, 1966).

Clark C. Abt, *A Strategy for Terminating a Nuclear War* (Boulder, Colo.: Westview Press, 1985).

C. Kenneth Allard, *Command, Control and the Common Defense* (New Haven, Conn.: Yale University Press, 1990).

Graham Allison, *The Essence of Decision: Explaining the Cuban Missile Crisis* (Boston, Mass.: Little, Brown, 1971).

Carl H. Amme, *NATO Without France* (Stanford, California: Hoover Institution, 1967).

William Arkin and Andrew Burrows, *British Nuclear Weapons in the Falklands* (Washington DC: IPS, 1984). Unpublished Paper.

The Atlantic Alliance Hearings, The Committee on Government Operations, US Senate, 89th Congress, Part 2, May 1966.

Authority to Order the Use of Nuclear Weapons, Congressional Research Service, House of Representatives, Subcommittee on International Security and Scientific Affairs of the Committee on International Relations, 1 December 1975, Washington DC.

Desmond Ball, *Can Nuclear War be Controlled?* (London: IISS, Adelphi Paper 169, 1981).

Desmond Ball, *Controlling Theatre Nuclear War* (Canberra, Australia: Australian National University, 1987). Working Paper 127.

Desmond Ball and Jeffrey Richelson (eds), *Strategic Nuclear Targeting* (Ithaca, NY: Cornell University Press, 1986).

Desmond Ball, *Targeting for Strategic Deterrence* (London: IISS, Summer 1983). Adelphi Paper No 185.

George Ball, *The Past Has Another Pattern* (London and New York: Norton Books, 1982).

James Bamford, *The Puzzle Palace: Inside the NSA* (Harmondsworth: Penguin, 1983).

Frank Barnaby (ed.) *Tactical Nuclear Weapons: European Persectives* SIPRI, (London: Taylor and Francis, 1978).

Paul Beaver, *Modern Royal Navy*, (Cambridge, UK: Patrick Stevens Publishers, 1982).

Francis Beer, *Integration and Disintegration in NATO* (Cincinatti: Ohio State University Press, 1969).

Coral Bell, *The Conventions of Crisis* (Oxford: Oxford University Press, 1971).

Richard K. Betts, *Soldiers, Statesmen and Cold War Crises* (Cambridge, Mass.: Harvard University Press, 1977).

Richard K. Betts, *Surprise Attack* (Washington DC: Brookings Institution, 1982).

J. D. Boutwell, P. Doty, and G. Treverton (eds), *The Nuclear Confrontation in Europe* (London & Sydney: Croom Helm, 1985).

Bruce Blair, *Strategic Command and Control* (Washington DC: Brookings Institution, 1985).

M. Blumenson and J. L. Stokesbury, *Masters of the Art of Command* (Boston, Mass.: Houghton Mifflin, 1975).

E. W. Boyd *et al.*, *NATO Management: Peace to Crisis Transition* (Santa Monica, Calif.: RAND Corporation, February, 1980). R-2576.

Paul Bracken, *The Command and Control of Nuclear Forces* (New Haven, Conn.: Yale University Press, 1983).

Lester H. Brune, *The Missile Crisis of October 1962* (Regina Books, 1986).

Alistair Buchan, *NATO in the 1960s* (London: Weidenfeld & Nicolson, 1960).

C. H. Builder, *Military Criteria for NATO TNF Options for the 1980s* (Santa Monica, Calif.: RAND Corporation, 1982). P-6772.

Rip Bulkley and Graham Spinardi, *Space Weapons: Deterrence or Delusion* (London: Polity Press, 1986).

Paul Buteux, *The Politics of Nuclear Consultation in NATO 1965–80* (Cambridge: Cambridge University Press, 1983).

C^3I Handbook Vol 1 (Palo Alto, California: EW Publications, 1986).

Dan Caldwell, *Permissive Action Links: A Description and a Proposal* Conference on Rethinking the Nuclear Weapons Dilemma in Europe, University of Minnesota, May 1985.

Duncan Campbell, *Unsinkable Aircraft Carrier* (London: Michael Joseph, 1985).

Duncan Campbell, *War Plan UK* (London: Paladin Books, 1982).

Carnegie Panel on US Security and the Future of Arms Control, *Challenges for US National Security*, Third Report 82-72322, 1982.

Ashton Carter, John Steinbruner and Charles Zracket (eds) *Managing Nuclear Operations* (Washington DC: Brookings Institution, 1987).

Center for Defense Information, *Command, Control and Communications Part I: The Command Structure and Early Warning System*, CDI Briefing Papers, 26 September 1986.

K. H. Cerny and H. W. Briefs (eds) *NATO in Quest of Cohesion* (Stanford, California: Hoover Institution, 1965).

Daniel Charles, *Nuclear Planning in NATO* (Cambridge, Mass.: Ballinger Press, 1985).

Steven Cimbala, *Strategic Nuclear War Termination* (New York: Praeger, 1986).

Harland Cleveland, *NATO: The Transatlantic Bargain* (New York: Harper & Row, 1970).

Thomas Cochran, William Arkin and Milton Hoenig, *Nuclear Weapons Databook Vol 1: US Nuclear Forces and Capabilities* (Cambridge, Mass.: Ballinger Press, 1984).

Command Structure of ACE, ACE Backgrounder Briefing, SHAPE Headquarters, B31-0787, 10 July 1987.

Communications and Crisis Management in the Alliance, WEU, 23rd Session, Part II, Document 757, 4 November 1977.

Conventional – Nuclear Operations, USACGSC RB100-30, vol. 1, US Army Command, 6 August 1976.
Robert Coulam and Richard Small (eds) *Advances in Information Processing in Organisations Vol 2* (New York: JAI Press, 1985).
Jonathan Crane, *Submarine* (London: BBC Books, 1984).
N. Crawford and A. Bloomgarden, *Soviet Military Aircraft* (Lexington, Mass.: Lexington Books, 1987).
Robert Crone, *NATO Nuclear Policy-Making* Unpublished Dissertation submitted in partial fulfilment of a PhD, University of California, Los Angeles, 1972.
John Cushman, *Command and Control of Theatre Forces: Adequacy* (Cambridge, Mass.: Harvard University, 1983).
Defense Advanced Research Projects Agency (DARPA), *A History of the ARPANET* (Washington DC: DARPA Publications, 1 April 1981).
Deputy Undersecretary of Defence for Command Control and Communications, *Modernisation of the WWMCCS Information System (WIS)*, Prepared for the Committee on Armed services, House of Representatives, No 96-916 and 97-333, 31 July 1982.
Guy De Carmoy, *The Foreign Policies of France 1944–68* (Chicago, Ill.: University of Chicago Press, 1970).
Andre De Staercke *et al.*, *NATO's Anxious Birth* (London: Hurst Press, 1985).
David Detzer, *The Brink: The Cuban Missile Crisis of 1962* (London: J. M. Dent, 1980).
Simon Duke, *US Defence Bases in the UK* (London: Macmillan, 1987).
Bjarne Eriksen, *The Committee System of the NATO Council* (Norway: Universitetforlaget, 1967).
First Use of Nuclear Weapons: Preserving Responsible Control, Hearing before the House Committee on International Relations, Subcommittee of International Security and Scientfic Affairs, 94th Congress, GPO, 1976.
Daniel Ford, *The Button* (New York: Simon & Schuster, 1987).
Lawrence Freedman, *Britain and Nuclear Weapons* (London: Macmillan, 1980).
Lawrence Freedman, *The Evolution of Nuclear Strategy* (London: Macmillan Press, 1982).
Terry Gander, *Encyclopaedia of the Modern British Army* (Cambridge, UK: Patrick Stevens Ltd Press, 1983).
Terry Gander, *The Modern Royal Air Force* (Cambridge, UK: Patrick Stevens Ltd Publications, 1984).
R. Girardet, *Problèmes Contemporains de Défense Nationale* (Paris: Dalloz, 1974).
Samuel Glasstone and Philip Dolan, *The Effects of Nuclear Weapons* (Tunbridge Wells, Kent: Castle House Press, 1980).
Kurt Gottfried and Bruce Blair (eds) *Crisis Stability and Nuclear War* (New York: Oxford University Press, 1988).
Shaun Gregory, *Command, Control, Communications and Intelligence in the Gulf War* (Canberra, Australia: Australian National University, Strategic and Defence Studies Centre, September 1991). Working Paper No 238.

F. Griffiths and J. C. Polanyi, *The Dangers of Nuclear War* (Toronto, Ont.: University of Toronto Press, 1979).
A. J. R. Groom, *British Thinking About Nuclear Weapons* (London: Frances Pinter, 1974).
Alfred Grosser, *The Western Alliance: European-American Relations Since 1945* (London: Macmillan, 1980).
Bill Gulley and Mary-Ellen Reese, *Breaking Cover* (New York: Simon & Schuster, 1980).
Alexander Haig, *Caveat* (London: Weidenfeld & Nicolson, 1984).
Max Hastings and Simon Jenkins, *The Battle For The Falklands* (London: Michael Joseph, 1983).
R. J. Hill, *Political Consultation in NATO* ORAE Memorandum No 64, Department of National Defence, Ottawa, Canada, March 1975.
Roger Hilsman, *To Move a Nation* (New York: Delta Books, 1967).
Malcolm Hoag, *Forward-Based Nuclear Systems in NATO in Historical Perspective: Lessons for SALT III* (Santa Monica, RAND Corporation, May 1980). P-6426.
Jolyon Howorth and Patricia Chilton (eds) *Defence and Dissent in Contemporary France* (London: Croom Helm, 1984).
Peter Jackson, *Strike Command* (London: Ian Allen, 1984).
Dorothy James, *The Contemporary Presidency* (Indianapolis: Bobbs-Merrill, 1974).
Jane's Military Communications 1989 (New York: Jane's Publications, 1989).
Jane's Weapons Systems 1987–88 (London and New York: Jane's Defence Publishers, 1988).
Jane's Weapons Systems 1988–89 (London and New York: Jane's Defence Publishers, 1989).
JAYCOR, *Theatre Nuclear Forces Survivability and Security: An Issue Evaluation Plan for Fixed Storage Sites*, DNA 5082F, Prepared for the Defense Nuclear Agency, October 1979.
Herman Kahn *et al.*, *War Termination: Issues and Concepts* (New York: Hudson Institute, 1968).
Marvin and Bernard Kalb, *Kissinger* (Boston, Mass.: Little, Brown, 1974).
Lawrence Kaplan, *NATO and the United States* (Boston, Mass.: Twayne Publications, 1988).
Henry Kissinger, *Years of Upheaval* (London: Weidenfeld & Nicolson, 1982).
Frank G. Klotz, *The US President and the Control of Strategic Nuclear Weapons* Unpublished PhD Thesis, Oxford University, 1980.
Klaus Knorr and Thornton Read, *Limited Strategic War* (New York: Praeger Press, 1962).
Louis W. Koenig, *The Chief Executive* (New York: HBJ Press, 1975).
W. L. Kohl, *French Nuclear Diplomacy* (Princeton, NJ.: Princeton University Press, 1971).
Roman Kolkowicz and Ellen Mickiewicz (eds) *The Soviet Calculus of Nuclear War* (Lexington, Mass.: Lexington Books, 1986).
Paul Lauren (ed.) *Diplomacy* (New York: Free Press, 1979).
Richard Ned Lebow, *Between Peace and War* (Baltimore, Md.: Johns Hopkins University Press, 1981).

J. M. Legge, *Theatre Nuclear Forces and the NATO Strategy of Flexible Response* (Santa Monica, Calif.: RAND Corporation, April 1983). R-2964-FF.

George F. Liska, *Nations in Alliance: The Limits of Interdependence* (Baltimore, Md.: Johns Hopkins Press, 1962).

Peter Malone, *The British Nuclear Deterrent* (London: Croom Helm, 1984).

Victor Marchetti and John D. Marks, *The CIA and the Cult of Intelligence* (New York: Dell Books, 1974).

Jeffrey P. McCausland, *Battlefield Nuclear Weapons and NATO Defence: Doctrines and Technologies*, Unpublished PhD Dissertation, Fletcher School of Diplomacy, 1984.

Jon McLin, *NATO and the Czechoslovakian Crisis, Part II: Invasion, Reaction and Stocktaking* Western Europe Series, vol. 4, 4, American University Field Staff, 1969.

The MITRE Corporation, *MITRE: The First Twenty Years* (Bedford, Mass.: MITRE Corporation, 1979).

Ben Moore, *NATO and the Future of Europe* (New York: Harper and Row, 1958).

Kenneth Myers (ed.) *NATO: The Next Thirty Years* (Boulder, Colo.: Westview Press, 1980).

NATO Final Communiques 1949–74 (Brussels: NATO Information Service, 1975).

NATO Final Communiques 1975–1980 (Brussels: NATO Information Service, 1981).

NATO Final Communiques 1981–1985 (Brussels: NATO Information Service, 1986).

NATO Final Communiques 1986–1990 (Brussels: NATO Information Service, 1991).

Richard M. Nixon, *The Memoirs of Richard Nixon (2 Vols)* (New York, USA: Warner Books, 1979).

The North Atlantic Treaty Organisation: Facts and Figures (Brussels: NATO Information Service, 1989).

George Orr, *Combat Operations C^3I: Fundamentals and Interactions* (Washington DC: Air University Press, 1983).

Robert E. Osgood, *NATO: The Entangling Alliance* (Chicago, Ill.: University of Chicago Press, 1962).

Robert E. Osgood, *Nuclear Control in NATO* (Washington DC: Washington Center for Foreign Policy Research, 1962).

T. W. Pakenham, *Naval Command and Control* (London: Brassey's Defence Publishers, 1989).

Richard Perle, *Transfer of Technology* Testimony to the Senate Subcommittee on Investigations of the Committee of Governmental Affairs, US Senate, 2nd Session, April 1984, 1422–54.

Andrew J. Pierre, *Nuclear Politics: The British Experience with an Independent Strategic Force, 1939–70* (Oxford: Oxford University Press, 1972).

R. L. Pfaltzgraff and U. Ra'anan (eds) *National Security Policy: The Decision-Making Process* (Hamden, Conn.: Archon Books, 1984).

John D. Read, *NATO's Theatre Nuclear Forces: A Coherent Strategy for the 1980s* (Washington DC: National Defense University Press, 1983).

Jeffrey T. Richelson and Desmond Ball, *The Ties that Bind* (London: Allen & Unwin, 1985).

Jeffrey T. Richelson, *The US Intelligence Community* (Cambridge, Mass.: Ballinger Press, 1985).

William H. Riker, *The Theory of Political Coalitions* (New Haven, Conn.: Yale University Press, 1962).

Maria Rita Saulle, *NATO and Its Activities: A Political and Juridicial Approach on Consultation* (New York: Oceana Publications, 1979).

Arthur Schlesinger, *A Thousand Days* (Boston, Mass.: Houghton Mifflin Books, 1965).

John Schlesinger, *Theatre Nuclear Force Posture in Europe* A Report to US Congress in Compliance with Public Law 93-365, 1974.

Barry Schneider, *Who Leads America If They Destroy Washington?* Washington DC National Institute for Public Policy, Information Series No 143, May 1983.

David N. Schwartz, *NATO's Nuclear Dilemmas* (Washington DC: Brookings Institution, 1983).

Seminar on Command, Control, Communications and Intelligence, Programme on Information Resource Policy (New Haven, Conn.: Harvard University, 1982).

Leon V. Sigal, *Nuclear Forces in Europe* (Washington DC: Brookings Institution, 1984).

J.M. Smith and C.P. Cotter, *The Powers of the President During Crises* (Washington DC: Public Affairs Press, 1960).

Glen H. Snyder, *Deterrence and Defense* (Princeton NJ.: Princeton University Press, 1961).

Malcolm Spaven, *Royal Navy Nuclear Capable Ships* Unpublished Briefing, ADIU, University of Sussex, 1985.

Paul Stares, *Command Performance: The Neglected Dimension of European Security* (Washington DC: Brookings Institution, 1991).

Peter Stein and Peter Feaver, *Assuring Control of Nuclear Weapons* CSIA Occasional Paper Series No 2 (Boston, Mass.: University Press of America, 1987).

Strategic Nuclear Weapons Policy United Kingdom House of Commons, Defence Committee, Session 1980–1.

Strategic Studies Center, *Increasing the Survivability of TNF Weapons in NATO*, SSC-TN-6070-2, April 1978.

R. Tourrain, *Rapport d'Information par la Commission de la Défense Nationale et des Forces Armées sur l'état et la modernisation des forces nucléaires françaises* No 1730, (Paris: Assemblée Nationale, October 1980).

US Security Issues in Europe: Burden Sharing and Offset, MFBR and Nuclear Weapons, Staff Report, Committee on Foreign Relations, US Senate, 93rd Congress, 2 December 1973.

USAF, *Digital European Backbone Facility*, FY1985, Military Construction Project Data, Project No 11470, 18 September 1984.

US Army, *Operations*, FM100-5, July 1976.

US Army, *The Pershing II Firing Battery*, FM 6-11, October 1984.

William Van Cleave and S.T. Cohen, *Tactical Nuclear Weapons: An Examination of the Issues* (New York: MacDonald and Jane's, 1979).

Martin Van Creveld, *Command in War* (Cambridge, Mass.: Harvard University Press, 1985).
Elliot Vandervanter, *Some Fundamentals of NATO Organisation* (Santa Monica, Calif.: Rand Corporation, 1963). RAND Memorandum RM-3559.
Arnold Wolfers, *Alliance Politics in the Cold War* (Baltimore, Md.: Johns Hopkins Press, 1959).
D. Yost, *France's Deterrent Posture and Security in Europe Part I: Capabilities and Doctrine*, Adelphi Paper 194, Winter 1984–5.
Charles A. Zracket, *The Impact of Command, Control Communications and Intelligence on Doctrine* (Bedford, Mass..: MITRE Corporation, November 1983).

Articles

'Advanced RITA for Hades missiles launch system', *Jane's Defence Weekly*, 10 October 1987, 801.
'Aerial Command Post', *Jane's Defence Weekly*, 13 February 1988, 253.
Francesco Amaduzzi, 'The Future of NATO Communications', *NATO Review* (January 1974) 14–18.
John Anderson, 'The Evolution of NICS', *Signal* (April 1979) 15–19.
John Anderson, 'NATO's Communications Requirement and the Role of NICSMA', *Journal of the Royal Signals Institute*, vol. 15, 4 (Spring 1982) 158–69.
W. D. Arkham, 'Southern Region C^3 Requirements', *NATO's Sixteen Nations* (June–July 1983) 82–4.
N. Ashford and J. Eisenhammer, 'A Cartesian guide to defending France', The *Independent*, 28 January 1988, 5.
'Avant-garde avionique', *Defence* (January 1988) 38–9.
David H. Bagley, 'C^3 in the NATO Theatre', *Signal*, March 1977, 8–11.
Desmond Ball, 'Nuclear War at Sea', *International Security*, vol. 10, 3 (Winter 1985) 3–31.
Leo Barker and John Dresser, 'The Initial Voice Switched Network: First Step Towards NICS', *Signal* (February 1976) 6–9.
Douglas Barrie, 'NATO Disbands ACCS Team', *Jane's Defence Weekly*, 19 August 1989, 287.
Douglas Barrie, 'NATO Rethinks Future of ACCS', *Jane's Defence Weekly*, 16 June 1990, 167.
Douglas Barrie, 'Towards NATO Software Standardisation', *Jane's Defence Weekly*, 19 August 1989, 308.
John Baylis, 'American Bases in Britain The Truman-Atlee Understandings', *The World Today* (August–September 1985) 155–9.
Richard K. Betts, 'Compound Deterrence vs. No First Use: What's Wrong is What's Right', *Orbis* (Winter 1985) 697–718.
Richard K. Betts, 'Surprise Attack: NATO's Political Vulnerability', *International Security*, vol. 5, 4 (Spring 1981) 117–49.
Kurt Birrenbach, 'Partnership and Consultation in NATO', *Atlantic Community Quarterly* (March 1964) 62–71.
Kent Black and Andrew Lindstrom, 'TACAMO', *Signal* (September 1978) 6–11.

Bruce Blair, *Arms Control Implications of ASW Programs*, US House of Representatives, Committee on International Relations, FY 1979, US GPO, 1978, 115–24.

David Boutacoff, 'WWMCCS Evolves to Meet Expanding Requirements', *Defense Electronics*, 4 November 1984, 79–85.

Jon L. Boyes, 'BICES: Linking NATO Intelligence', *Signal* (October 1987) 19–20.

Paul Bracken, 'Collateral Damage and Theatre Warfare', *Survival*, vol. 22, 5 (Sept–Oct 1980) 203–7.

Paul Bracken, 'Command and Control for a Long War', *Air Force Magazine*, April 1980, 50–4.

Paul Bracken, 'The NATO Defense Problem', *Orbis*. (Spring 1983) 83–105.

Paul Bracken, 'The Political Command and Control of Nuclear Forces', *Defense Analysis*, vol. 2, 1 (1986) 11–20.

Paul Bracken, 'Urban Sprawl and NATO Defence', *Survival*, vol. 18, 6 (November–December 1976) 254–60.

Michael Brecher, 'Towards a Theory of International Crisis Behaviour', *International Studies Quarterly*, vol. 21, 1 (1977) 39–74.

Ronald F. Brender *et al.*, 'What is ADA?', *IEEE Computer* (June 1981) 17–24

'Britain's Fence Takes Shape', *Jane's Defence Weekly*, 15 December 1984, 1078–80.

'British Military Communications: Part 1', *Defence Materiel* (March–April 1980) 34–47.

'British Military Communications: Part 2', *Defence Materiel* (May–June 1980) 94–108.

'British Military Communications: Part 3', *Defence Materiel* (July–August 1980) 114–23.

Stephen Broadbent, 'Joint-STARS: Force Multiplier for Europe', *Jane's Defence Weekly*, 18 April 1987, 729–33.

Paul Brown and David Fairhall, 'Anger Over Plan For Radar Station at Beauty Spot', The *Guardian*, 5 April 1990, 24.

Paul Brown, 'MoD Digs Deep for Bunker', The *Guardian*, 25 July 1985, 4.

Alistair Buchan, 'The Reform of NATO', *Foreign Affairs*, vol. 40, 2 (January 1962) 27–43.

John C. Burney, 'Nuclear Sharing in NATO', *Military Review*, June 1969, 62–8.

Duncan Campbell, 'The Deterrent Goes to War', *New Statesman*, 1 May 1981, 9–11.

Duncan Campbell, 'Maggie's Bunker', *New Statesman*, 24 September 1982, 6–8.

Duncan Campbell, 'New Spy Station to Look Inside USSR', *New Statesman*, 21–28 December 1984, 6.

Duncan Campbell and Patrick Forbes, 'New Whitehall Bunker', *New Statesman*, 26 July 1985, 6.

Duncan Campbell, 'The Parliamentary Bypass Operation', *New Statesman*, 23 January 1987, 8–10.

Duncan Campbell, 'Too Few Bombs to Go Round', *New Statesman*, 29 November 1985, 10–12.

James W. Canan, 'Increasingly Interoperable NATO', *Air Force Magazine*, June 1985, 97–103.

Steven Canby, 'NATO Strategy: Political-Military Problems or Divergent Interests and Operational Concepts', *Military Review*, April 1979, 50–8.

Ashton B. Carter, 'The Command and Control of Nuclear War', *Scientific American*, vol. 252, 1 (January 1985) 20–7.

Tony Catterall, 'War Game "Idiocy" fires Kohl's Resolve', The *Observer*, 30 April 1989, 27.

'Cemetery Net Upgrade (EUCOM C3)', House of Representatives, FY1981, Appropriations Committee, DOD Authorisations, 12 September 1979, 748.

Daniel Charles, 'Who Controls NATO's Nuclear Weapons', *Bulletin of the Atomic Scientists*, (April 1985) 45–8.

John Cittadino, 'NATO's Tri-Service Group on Communications and Electronic Equipment', *Signal* (October 1984) 61–76.

John J. Clark, 'Is the NATO Alliance Structure Appropriate for the 1980s?', *Military Review*, (April 1979) 25–34.

John Clark, 'NATO Standardisation: Panacea or Plague?', *Military Review* (April 1979) 59–65.

Conférence de presse de Monsieur François Mitterrand à l'issue des manoeuvres Kecker Spatz, Présidence de la République, 24 September 1987.

'Conflicting Systems Negate NATO C^3 Lead', *C^3I Report*, 25 January 1988, 7.

Anthony H. Cordesman, 'NATO C^3I: The Problem of Nuclear War', *Armed Forces Journal International*, (December 1982) 64.

Craig Covault, 'New Missile Warning Satellite to be Launched on First Titan 4', *Aviation Week and Space Technology* (20 February 1989) 34–40.

'Crisis Cabinet for NATO', *Atlantic Community News* (March 1980) 2.

J. Critchley *et al.*, 'Can the Use of Nuclear Weapons be Decided Upon Collectively?', *NATO Letter* (October 1965) 18–22.

A. J. Culligan, 'An Overview of EMP Effects and their Control', *Journal of the Society of Environmental Engineers* (September 1985) 9–11.

Katie Cutler, 'Inside the Looking Glass', *Airman* (September 1982) 7–12.

P. Denarie, 'Helios and Syracuse: military space at technospace exhibition', *Defense & Armament Héraclès International*, 81 (February 1989) 57–8.

'Der Iwan Kommt – und Feste Druff', *Der Spiegel*, 18 (1989) 23–7.

Francois De Rose, 'Inflexible Response', *Foreign Affairs*, vol. 61, 1 (Fall 1982) 136–50.

Tony Devereux, 'OTH Radar Comes of Age', *Defence Attaché*, 3 (1987) 34–7.

J. Deygout, 'RITA: a Modern Response to Battlefield Communications Needs', *Signal* (March 1984) 26–33.

Loren Diedrichsen, 'Accomplishments in Communications Interoperability Amongst NATO Tactical Forces', *Easion* (1978) 28–31.

Michael Donnheim, 'E-6A TACAMO Tail Damage Prompts Delvery Delay', *Aviation Week and Space Technology* (17 April 1989) 28–9.

Joseph D. Douglass, 'Soviet Nuclear Strategy in Europe: A Selective Targeting Doctrine?', *Strategic Review* (Fall 1977) 12–23.

M. Ducal, 'L'Arme Nucléaire: pour quoi faire?', *Défense Nationale* (February 1989) 13–29.

W.F. Eekelen, 'Development of NATO's Nuclear Consultation', *NATO Letter*, 7 (July 1970) 2–6.

'Ellsberg: Who has Finger on US Nuclear Trigger', *Omaha World Herald* (30 August 1981) 1.

M. Faivre, 'Le budget de la défense pour 1989', *Défense Nationale* (January 1989) 173
M. Faivre, 'Le renseignement du champ de bataille', *Défense Nationale* (August–September 1988) 174–8
M. Faivre, 'Menaces et risques majeurs pour la sécurité de la France', *Défense Nationale* (December 1988) 20–4.
P. M. Ford, 'The Defence Communications Network', *Journal of the Royal Signals Institute* (Summer 1989) 68–74.
G. D. Foster, 'Contemporary Command and Control Theory and Research: The Failed Quest for a Philosophy of Command', *Defense Analysis*, vol. 4, 3 (September 1988) 201–28.
Nancy Foster, 'Citizens Jam Nuclear Radio Network', *Bulletin of the Atomic Scientists* (November 1988) 21–5.
'France and Germany agree to set up 'hot-line'', *Christian Science Monitor* (26 August 1985) 2.
'France buys fourth E-3', *Defence* (December 1987) 606.
J. A. D. Francis, 'The NATO Integrated Communications System', *Journal of the Royal Signals Institute*, vol. 13, 5 (Summer 1978) 17–23.
Franco-German summit statement, 27–8 February 1986 (exerpts), translated and reproduced in *Survival*, vol. 28, 4 (July–August 1986) 366–7.
Lawrence Freedman, 'British Nuclear Targeting', *Defense Analysis*, vol. 1, 2 (1985) 81–99.
R. N. B. Freeman, 'The NATO Integrated Communications System – Stage I', *Journal of the Royal Signals Institute*, vol. 17, 3 (Winter 1985) 156–9.
Simon Freeman and Barrie Penrose, 'How Britain Would Have Coped if the Cabinet Had Been Killed', *The Times* (21 October 1984) 10.
'The French Armaments Industry in 1989', *Defence* (March 1989) 187–212.
'French nuclear command post enters operational service', *Aviation Week and Space Technology* (8 February 1988) 253.
Alton Frye, 'Nuclear Weapons in Europe: No Exit from Ambivalence', *Survival* (May 1980) 98–106.
'The future of French aerospace', *Defense & Armament Héraclès International*, 77 (October 1988) 51–2.
P. M. Gallois, 'New Teeth for NATO', *Foreign Affairs*, vol. 39, 4 (October 1960) 67–80.
John R. Galvin, 'The Challenges in Command and Control in USEUCOM', *Signal* (March 1988) 25–8.
John R. Galvin, 'European Security: A Military Perspective', *RUSI Journal* (Autumn 1991) 5–9.
David Gates, 'American Strategic bases in Britain: The Agreements Governing their Use', *Comparative Strategy*, vol. 8 (1989) 99–123.
Thomas E. Genon, 'Automating NATO's Command Centres', *Signal* (December 1982) 45–51.
Colin S. Gray, 'NATO's Nuclear Dilemma', *Policy Review* (Fall 1982) 97–116.
Colin S. Gray, 'Theatre Nuclear Weapons: Doctrine and Postures', *World Politics* (January 1976) 300–14.
Shaun Gregory, 'French Nuclear Command and Control', *Defence Analysis*, vol. 6, 1 (1990) 49–69.

Everett D. Grienke, 'C^{3}I In Allied Command Europe', *Signal* (December 1983) 81–3.
Gary L. Guertner, 'Flexible Response in NATO Military Strategy: Deterrent or Escalation Trap', *Comparative Strategy* (May 1989) 335–51.
Gary L. Guertner, 'Nuclear War in Suburbia', *Orbis*, vol. 26, 1 (1982) 49–70.
John Guthrie, 'The Doomsday Trigger', *International Combat Arms* (July 1986) 92–3.
John Guthrie, 'Underground Command Posts', *International Combat Arms* (May 1987) 16–21.
R. T. Hanson and K. Straussman, 'The US Armed Forces European Telephone System', *Signal* (October 1986) 61–3.
P. Harding, 'C^{3}I Supporting the Commanders Game Plan', *Signal* (October 1987) 27–34.
R. B. Harrison, 'WWMCCS ADP: The Performance Story', *Signal* (January 1985) 47–53.
Douglas M. Hart, 'Soviet Approaches to Crisis Management: the Military Dimensions', *Survival*, vol. 26, 5 (September–October 1984) 214–23.
Jakob Hartog, 'ACCS: A Challenge and an Opportunity', *Signal* (December 1983) 72–4.
Jakob Hartog, 'The Development of an Air Command and Control System for NATO's Air Forces in Europe', *NATO's Sixteen Nations* (October 1985) 76–8.
Jakob Hartog, 'The Development of a NATO Air Command and Control System', *Signal* (October 1984) 51–5.
P. Henin, 'Les télécommunications militaires', *Défense Nationale* (May 1987) 143–55.
C. Hernu, 'Répondre aux défis d'un monde dangereux', *Défense Nationale* (December 1981) 44–53.
Mark Hewish, 'Britain's Air Defence Grows New Teeth', *New Scientist* (12 March 1981) 682–4.
Mark Hewish, 'British Developments in Satellite Communications', *International Defence Review*, 5 (1985) 721–3.
Irving Heymont, 'The NATO Bilateral Forces', *Orbis*, vol. 9, 4 (1966) 1025–41.
Peter Hill-Norton, 'Crisis Management', *NATO Review* (October 1976) 6–9.
G.D. Hingorami, 'Digital Replacement of the ACE-HIGH Transmission System', *Signal* (December 1980) 61–4.
G. D. Hingorami and R. Brand, 'Architectural Framework for the Evolution of the NATO Integrated Communications Systems', *Signal* (October 1985) 55–68.
A.M.E. Hoare, 'Project Fastnet – The Modernisation of the Army Fixed Telecommunications System in the UK', *Journal of the Royal Signals Institute*, vol. XVIII, 3 (Winter 1987) 160–6.
Arthur Hockaday, 'Nuclear Management in NATO', *NATO Letter*, (May 1967) 3–7.
Stanley Hoffmann, 'NATO and Nuclear Weapons: Reason and Unreason', *Foreign Affairs* (Winter 1981–1982) 327–46.
Chet Holifield, 'Nuclear Control in NATO', *Atlantic Community Quarterly* (Winter 1965–66) 496–8.

John W. Holmes, 'Fearful Symmetry: the Dilemmas of Consultation and Co-ordination in the North Atlantic Treaty Organisation', *International Organisations*, vol. 22 (1968) 821–40.

Johan J. Holst, *Some Reflections on Alliance Strategy and the Problem of Command, Control and Communications*, Paper presented to the Conference on Coalition Defence and C^3I, MIT and MITRE Corporation, 13–14 October 1983.

Ted Hooton, 'NADGE Enhancement Lay Ground Work for ACCS', *International Defence Review*, 7 (1988) 781–3.

Thomas Horton, 'American ADA busy in Europe', *Jane's Defence Weekly* (24 November 1984) 954.

Fred Ikle, 'NATO's First Nuclear Use: A Deepening Trap', *Strategic Review* (Winter 1980) 18–23.

A. Nejt Ime, 'Electronic Warfare and NATO Communications', *Signal* (March 1978) 66–73.

'Inside the Elysée', The *Sunday Times* (Magazine), 30 July 1985, 27–30

'Interview with General Bernard Rogers', *US News and World Report* (15 June 1981) 25.

J. Isnard, 'France and Europe's Security', *Defense & Armament Héraclès International* (November 1987) 71.

J. Isnard, 'France to update nuclear arsenal with ASMP', *Jane's Defence Weekly* (30 July 1988) 6.

J. Isnard, 'French looking at land-based M-4', *Jane's Defence Weekly* (29 October 1988) 1050.

J. Isnard, 'Le code d'engagement de la force nucléaire', *Le Monde* (20 May 1981) 12.

J. Isnard, 'Pure deterrence - or tainted strategy', *Defense & Armament Héraclès International* (June 1988) 31.

Joachim Jaenicke, 'Political Consultation – Reality or Formality', *NATO Letter* (April 1969) 15–17.

'Jaguar Mission', *RAF News* (9–23 September 1983) 11.

Peter Jenner, 'NATO-wide Communications Improved', *NATO Letter* (April 1970) 7–8.

Stephen L. Johnston, 'Soviet Electronic Warfare: A Review of Published Material', *International Defence Review*, 12 (December 1985) 9–14.

Joe Jones, 'Defense Communications System AUTODIN', *Signal* (September 1969) 74–5.

Robert A, Leahy, 'C^3 for Strategic Nuclear War Termination', *Signal* (August 1988) 38–42.

Philip Karber, 'Nuclear Weapons and Flexible Response', *Orbis* (Summer 1970) 284–97.

M. King and P. B. Fleming, 'An Overview of the Effects of Nuclear Weapons on Communications Capabilities', *Signal* (January 1990) 59–68.

Frederic L. Kirgus, 'NATO Consultations as a Component of National Decision-Making', *American Journal of International Law*, vol. 73, 3 (July 1979) 372–406.

Harold A. Kissinger, 'Organisation for Crisis Management in NATO', *Signal* (March 1975) 66–70.

Klaus Knorr, 'Nuclear Weapons: "Haves" and "Have-Nots" ', *Foreign Affairs*, vol. 36, 1 (April 1957) 167–78.
D. Kyle and D. Meyer, 'Has NATO C3I Gone Haywire?', *Armed Forces Journal International* (December 1982) 56–75.
J. Lacaze, 'La politique militaire', *Défense Nationale*, (November 1981) 7–26.
'L'engagement nucléaire ne peut reposer que sur la décision d'un seul', *Le Monde* (19 November 1980) 12.
J. M. Leonorvitz, 'Attitude lost on France's Telecom 1B', *Aviation Week and Space Technology* (25 January 1988) 30.
Eric J. Lerner, 'Tactical C^3: Survival Under Duress', *IEEE Spectrum*, (October 1982) 56–7.
'Les Etats-Unis acceptant de vendre à la France des émetteurs pour sa force dissuasion', *Le Monde* (16 September 1982) 14.
'L'espace pour voir, écouter, communiquer', *Défense Nationale* (July 1986) 140–50.
Joris Lok, 'NATO Studies Lack of Tactical Net', *Jane's Defence Weekly* (20 January 1990) 253.
Karl Lowe and Thomas Durrell Young, 'Multinational Corps in NATO', *Survival* (January–February 1991) 66–77.
Hugh Lucas, 'Ptarmigan, £3.1 billion more, loses to RITA', *Jane's Defence Weekly* (16 November 1985) 1066.
Hugh Lucas, 'US Radio Link to Submarines at 400ft Depth in Mediterranean', *Jane's Defence Weekly* (21 July 1984) 52.
William G. Maclaren, 'The Force Multiplier: C^3 in NATO', *NATO's Sixteen Nations*, (April–May 1983) 16–20.
P. R. Mallone, 'Command and Control: Some Ace Aspects', *NATO's Fifteen Nations* (October–November 1979) 24–30.
Geoffrey Manners, 'NATO's Aerial Sentries on Skywatch Duty', *Jane's Defence Weekly* (24 March 1984) 452–3.
J. B. Margeride, 'Réflexions sur les systèmes à l'ère électro-informatique', *Stratégique* (Winter 1984) 115–17
I. Margine, 'L'Avenir de la disuassion', *Defénse Nationale* (April 1978) 7–29.
Dennis C. Marquis, 'New Directions in NATO C^3', *Signal* (December 1982) 39–42.
'Martello – Modern 3-D Surveilllance Radar', *GEC Journal of Research*, vol. 3, 2 (1985) 104–13.
J. J. Martin, 'Nuclar Weapons in NATO's Deterrent Strategy', *Orbis* (Winter 1979) 875–95.
Laurence Martin, 'Theatre Nuclear Weapons and Europe', *Survival* (November 1974) 268–76.
F. Maurin, 'Entretien avec le Général Maurin', *Défense Nationale* (July 1974) 17.
'M. Mauroy: un ordre d'engagement de l'arme nucléaire est peu compatible avec une concertation préalable', *Le Monde* (11 March 1984) 11.
Clarence E. McKnight, 'The Army's National Role in NATO Communications', *Signal* (December 1980) 43–50.
A. L. Meier, 'BICES: A Central Region Perspective', *International Defence Review*, 10 (1986) 1445–9.

'The Men Who Wait for Nuclear Attack', The *Guardian* (27 October 1980) 4.
Stewart Menaul, 'Great Britain and NATO Theatre Nuclear Forces', *Strategic Review* (Spring 1981) 61–6.
F. O. Miksche, 'The Case for Nuclear Sharing', *Orbis* (Fall 1961) 292–305.
John Morgenstern, 'C^3 for Tactical Nuclear Forces in Europe', *Signal* (December 1980) 57–8.
John Morgenstern, 'Strategic and Theatre Command and Control Systems', *Signal* (November–December, 1978) 50–5.
Frederick Mulley, 'NATO's Nuclear Problems: Control or Consultation?', *Orbis* (Spring 1964) 20–8.
'NATO Looks Towards 2008', *Defence Electronics and Computing* (supplement to *International Defence Review*), 12 (1989) 151–3.
'NATO Set to Upgrade SHAPE System', *Jane's Defence Weekly* (October 1987) 16.
'NATO's Shelterised HF Communications System', *Jane's Defence Review*, vol. 3, 1 (1982) 85.
Richard Neff, 'NATO Political Consultation: Fact or Myth?', *NATO Review*, vol. 23, 1 (January 1975) 7–9.
'The Nerve Centre of United States Nuclear Strategy', *NATO Letter* (May 1966) 19–20.
'New AD Sensors for the UK', *Defence Attaché*, 2 (1985) 6.
'New HF radio links for French navy', *Maritime Defence*, (September 1986) 11.
'The New NADGE', *International Defence Review*, 5 (1985) 755–8.
'New NATO C^3 System', *Defence* (August 1987) 7.
Sam Nunn, 'NATO Strategy', *Survival*, (January–February 1977) 30–8.
Peter Osnos, 'British say they can veto cruise missile firing', *The Washington Post* (1 November 1983) 1.
Frank A. Partlow, 'The NATO Military Committee and the International Military Staff: Some Rationales and a Proposal for Re-organisation', *RUSI Journal*, vol. 119 (September 1974) 11–18.
Rupert Pengelley, 'HEROS and Wavell: Battlefield ADP Enters a NEW Era', *International Defence Review*, 10 (1986) 1459–64.
R. Pengelley, 'OPCON 2 – a Strategic maritime C2 System', *Defence Attaché*, 6 (1981) 9.
Richard Perle, Testimony, *US Senate Committee on Armed Services 1985*, US GPO, 1985, Pt VII, 157–163.
Tekla S. Perry, 'NATO Credibility: An Expert Assessment', *IEEE Spectrum* (October 1982) 71–5.
Walter Pincus, 'President's Command Jet Shifted Inland', *The Washington Post* (23 September 1983) 3.
James Polk, 'The Realities of Tactical Nuclear Warfare', *Orbis* (Summer 1973) 439–47.
A. J. Poppleton, 'The Survivability of Battlefield Communications', *Journal of the Royal Signals Institute* (Spring 1985) 43–6.
Barry Posen, 'Inadvertant Nuclear War?: Escalation and NATO's Northern Flank', *International Security*, vol. 7, 2 (Fall 1982) 28–54.
'RAF Developments',*Defence Attaché*, 3 (1981) 30.
'RAF Honington', *RAF News* (16 December 1982) 12–13.

Bob Raggett, 'NATO C^3 in Disarray', *Jane's Defence Weekly* (8 September 1984) 418–20.
R.J. Raggett, 'Naval Communications Experience – the Good and Bad Aspects of Technology', *Maritime Defence* (August 1979) 312–13.
'RAMSES communications network completed', *Maritime Defence* (May 1988) 196.
'RAMSES nuclear order network delivered', *Jane's Defence Weekly* (30 April 1988) 849.
George Rathjens, 'Flexible Response Options', *Orbis* (Fall 1974) 677–88.
C.J. Rauch and W.E. Edgington, 'NATO Develops a Long-Term Air Command and Control Plan', *Defence Systems Review* (January 1984) 12–17.
'Raytheon to Install Phased-Array radar at BMEWS Site in England', *Aviation Week and Space Technology* (12 September 1988) 131.
Jeffrey Record, 'To Nuke or Not to Nuke: A Critique of Rationales for a Tactical Nuclear Defense of Europe', *Military Review* (October 1974) 3–13.
'Regency Net Contract', *Signal*, (April 1984) 72.
Richard Reaser and Robert Wood, 'AUTODIN is On', *Signal* (April 1970) 6–9.
'Revealed, US Role in UK SSBN Communications Link', *Defence* 1 (January 1987) 6.
D. Richardson, 'Lance into Battle', *Flight International* (30 April 1977) 1192–95.
Thomas M. Rienzi and David Coleman, 'The Transition in NATO Telcommunications', *Signal* (September 1979) 32–5.
Bernard Rogers, 'Greater Flexibility for Flexible Response', *Strategic Review* (Spring 1983) 11–19.
Joseph Rose, 'AUTOVON and AUTODIN: Past Present and Future', *Signal* (April 1969) 12–15.
Scott D. Sagan, 'Nuclear Alerts and Crisis Management', *International Security*, vol. 9 (Spring 1985) 99–139.
Benjamin Schemmer, 'Haig Says NATO Can Now Expect 8–14 Days Warning, Not 48 Hours', *Armed Forces Journal International* (October 1977) 16.
Benjamin Schemmer, 'Who Would Start a War?', *Armed Forces Journal International* (December 1981) 36–46.
M. Schneider, 'Une monarchie nucléaire', *Le Monde* (8 December 1983) 13.
Daniel Schulman, 'Nuclear Strategy and the Problem of Command and Control', *Survival*, vol. XXIX 4 (July–August 1987) 336–59.
James Schultz, 'Northern Europe to Receive New Battle Management System', *Defense Electronics* (May 1985) 64–8.
Franz-Jospeh Schulze, 'Command, Control and Communications in NATO', in W. Kaltefleiter and Ulrike Schumacher (eds) *Conflicts, Options, Strategies in a Threatening World*, Christian Albrechts University (Kiel: Institute of Political Science, 1981) 207–34.
A. Schwartzbrod and M. J. Gething, 'French air power – the turning point', *Defence* (September 1988) 642–6.
A. Schwartzbrod, 'French Navy in the Gulf', *Defence* (September 1987) 505.
Herbert Scoville, 'Flexible Madness', *Foreign Policy* (Spring 1974) 164–77.

'Shackleton is Better than Nothing says C-in-C', *Jane's Defence Weekly* (30 November 1985) 1182–3.

'SHAPE Communications', *Signal* (December 1959) 7–8.

S. Shiveley, 'A Local Area Network for NATO Headquarters', *Signal* (May 1984) 137–52.

Anne Sington, 'French Link in a NATO Communications System', *NATO Letter* (April 1965) 10–17.

Anne Sington, 'NATO Satcom', *NATO Letter* (March 1968) 11–16.

'SINTAC – the French answer to JTIDS', *Defence* (December 1987) 630–1.

Eberhard Skibbe, 'NATO Command and Control Systems', *Wehrtechnik* (June 1984) 82–6.

P.J. Sloyan, 'America in secret computer deal with France', The *Guardian* (5 February 1985) 11.

Leon Smith, 'C^3I – The New European Systems', *Armed Forces Journal International* (February 1984) 54–60.

R. Smith, 'Fingers on the French Button', The *Observer* (3 July 1983) 6.

J. Sochaczewski, 'NATO Command and Control and Information Systems', *Military Technology* 10 (October 1988) 31–7.

J.M. Sochaczewski, 'The Role of Communications in NATO', *Military Technology*, 6 (1984) 154.

R.A. Spackman, 'Lance: Field Artillery Missile System', *Journal of the Royal Artillery* (September 1972) 127–33.

Malcolm Spaven, 'Communicating with Submarines', *Jane's Defence Weekly* (23 November 1985) 1152–6.

Malcolm Spaven, 'ELF – Surviving the Traumas Part II', *Jane's Defence Weekly* (30 November 1985) 1194–7.

Timothy Stanley, 'Decentralising Nuclear Control in NATO', *Orbis* (Spring 1963) 41–8.

Timothy Stanley, 'NATO's Strategic Doctrine', *Survival* (November 1969) 342–9.

John Steinbruner, 'National Security and the Concept of Strategic Stability', *Journal of Conflict Resolution*, vol. 22, 3 (1978) 411–28.

John Steinbruner, 'Nuclear Decapitation', *Foreign Policy* (Winter 1981–2) 16–28.

Johannes Steinhoff, 'Strategy and Stability', *NATO's Fifteen Nations* (April–May 1967) 27.

'The Syracuse II programme', *Defense & Armament Héraclès International* (December 1988) 72.

'Test alert attributed to Gates', *The Washington Post* (May 17 1960) 1.

Greg Todd, 'C^1 Catharsis', *Defense Analysis*, vol. 4, 3 (September 1988) 203.

'The Transfer of SHAPE to Belgium', *NATO Letter* (December 1966) 12–13.

G. Turbe, 'The French Armed Forces at the Turn of the Century', *International Defence Review*, 3 (1988) 235–6.

G. Turbe, 'Testing Franco-German Military Co-operation', *International Defence Review*, 11 (1987) 1458.

'Two Important Naval Communications Contracts', *Defence Materiel* (January–February 1985) 19.

'UK Military Satellite Communications', *Military Technology*, Electronics Special Issue 1 (1984) 87–92.

'UK to update Missile Detection System', *Jane's Defence Weekly* (7 June 1986) 1028.

R. Ullman, 'The Covert French Connection', *Foreign Policy* (Summer 1989) 3–33.

Mark Urban, 'Fibre Optics are Arteries of Information Network', The *Independent* (12 January 1989) 5.

Hans Van Gelder, 'Improving the Survivability of NATO Communications', *Signal* (May 1983) 145–55.

Karl H. Wagner, 'Interoperability in Command and Control by Networks', *NATO's Fifteen Nations* (October–November 1979) 49–55.

Brian Wanstall, 'Closing NATO's Backdoor', *Interavia*, vol. 2 (1983) 125–7.

Larry K. Wenz, 'NATO/National Strategic Defense Communications Systems: Interoperability and Rationalisation – Rhetoric or Reality', *Signal* (October 1985) 83–90.

Larry K. Wentz, 'Outlook for NATO communications', *Signal* (December 1982) 53–60.

Thomas C. Wiegele, 'Nuclear Consultation Processes in NATO', *Orbis* (Summer 1972) 462–87.

David Williams, 'NATO Merges Planners, Procurers to End Communications Patchwork', *Defense News* (19 May 1986) 4.

Phil Williams, 'NATO Crisis Management: Dilemmas and Trade-Offs', *The Washington Quarterly*, vol. 12, 2 (Spring 1989) 29–39.

Manfred Woerner, 'NATO Defences and Tactical Nuclear Weapons', *Strategic Review* (Fall 1977) 11–18.

Wilfried Wohl, 'Nuclear Strategy in NATO and the MLF', *Political Science Quarterly*, vol. LXXX, 1 (March 1965) 89–109.

Albert Wohlstetter, 'Nuclear Sharing: NATO and the N + 1 Country', *Foreign Affairs*, vol. 39, 3 (April 1961) 355–87.

D. Wood, 'Control of the Eastern Seaways', *Interavia*, 1 (1980) 44.

Gary Yerkey, 'Greek Participation Will Aid NATO Surveillance Program', *Christian Science Monitor* (17 September 1985) 13.

Kenneth F. Zitzman, 'International System Problems as Exemplified by NATO Project ACE-HIGH', *Signal* (November 1960) 44–7.

Index